D0769532

Discovering
ANTIQUE MAPS

Alan G. Hodgkiss

Shire Publications Ltd.

CONTENTS

Introduction 4

The origins and development of mapmaking 15

The printed map 19

British regional mapping: 1579-1700 21

British regional mapping: 1700-1860 50

The formative years of the Ordnance Survey 64

Suggestions for further reading 67

Where to obtain and consult early maps 68

Index 72

1. Scales of distance on Jaillot's map of Hungary, 1696.

Copyright © 1971 and 1981 by A. G. Hodgkiss. Editions in 1971, 1975, 1977 and 1981, fourth edition reprinted 1983, 1986, 1988 and 1992. Number 98 in the Discovering series. ISBN 0 85263 581 8.

Printed in Great Britain by C. I. Thomas & Sons (Haverfordwest) Ltd, Press Buildings, Merlins Bridge, Haverfordwest, Dyfed SA61 1XF.

'What greater pleasure can there now bee, than to view those elaborate maps of Ortelius, Mercator, etc. . . . Me thinkes it would well please any man to look upon a Geographical map . . . to behold as it were all the remote Provinces, Townes, Citties of the World.'

Anatomy of Melancholy, Robert Burton, 1621

It is hard to believe that there is anyone who is unmoved by early maps—whose interest remains unstirred by a Speed county map, or whose imagination fails to respond to a 17th century map of far-off lands. There are certainly many people who like to have a framed antique map on the wall of their living-room simply for its decorative qualities and early maps are undoubtedly very attractive when displayed in this way. They have, however, so much more to offer than mere decoration. They epitomise the geographical knowledge and the customs of their day; their ornamentation reflects contemporary artistic style and taste; their colouring is often brilliant; the development of the techniques by which they were made is full of interest; so too are the mapmakers themselves, men such as Mercator—geographer, instrument maker, engraver and scribe, at one time accused of being a heretic, inventor of the projection which bears his name and the man who used the word 'atlas' for the first time to denote a bound collection of maps—or John Ogilby, who introduced the still popular 'strip-map' method of depicting roads in 1675, and whose varied career and talents made him one of the more colourful personalities in British mapmaking.

If this small book can do anything towards persuading an unbeliever of the attractions of early maps or, alternatively, add to the enjoyment of the converted it will have achieved its purpose.

OCEANUS

Swash lettering— the flourishing lettering often used to fill in unwanted space.

INTRODUCTION

Mapmaking, or cartography, has a long, many-faceted history, the whole field of which could not be covered in a large book, much less a short one and the present work can attempt only to provide the reader with an introduction to the subject and, it is hoped, encourage him to undertake wider investigations. The opening section will consider the terminology of early maps and their constituent features, followed by a résumé of mapmaking to the end of the 16th century. The remainder of the book will concentrate on the development of British regional mapping from the time of Saxton onwards.

What is meant by an 'original' map?

People are sometimes concerned as to how they can distinguish an original map from a modern reproduction, so it is appropriate that first something should be said about the term 'original' as applied to maps. Before the invention of printing, maps were drawn on vellum, sheepskin or other suitable material and could only be reproduced by laborious hand-copying. Further classes of map, e.g. the estate map, are produced in manuscript only, with no printing process involved. So far the word 'original' can be applied in the same way as it would to a painting, in the sense of an individual piece of artwork. The advent of printing, however, allowed the production of multiple copies and when we are talking about printed maps the word 'original' means a **print** or **impression** taken from a wood block or from a copper or steel plate. This is what we normally mean by an 'original' map—a print and not an individual piece of artwork.

The earliest printed maps were impressions taken from engravings on wood blocks; a **relief** process in which detail and lettering were left standing to form a printing surface and areas without detail were cut away. The next development, that of engraving on sheets of copper, brought greater flexibility and precision into mapmaking, so that lettering became more flowing and detail could be finer and neater. Copper engraving is an **intaglio** process, the detail being cut into the copper sheet. This is then inked, wiped over leaving ink only in the cuts, and finally an impression taken by passing through a printing press along with a dampened sheet of paper. Only in the 19th century was it challenged by engraving on steel, a process which did not call for any new production methods but, because of the hardness of the metal, allowed many more impressions to be taken before the plate became worn.

Lithographic printing also came into being in the 19th century and is a surface or **planographic** process depending, on the one hand on the antipathy between grease and water and, on the other, on the affinity between one greasy substance and another. As the term **lithography** implies the design was originally drawn with greasy ink on a specially prepared stone from which impressions were then taken. Nowadays, metal, plastic and paper plates have superseded stone and the image is now transferred photographically on to the plate, from which copies can be run off at speed on offset-litho printing machines.

Distinguishing an original printed map from a modern facsimile reproduction should present no problem for, in most cases, the name of the publisher of the facsimile will be clearly stated outside the border of the map. If any doubt exists it is helpful to look for the impression mark outside the border of an original map made by the press squeezing the printing plate into the paper—a modern litho-printed reproduction will have no such mark. There is no intent to make forgeries when producing facsimile maps, indeed they form a valuable service to serious students and map enthusiasts and on the whole they are of excellent quality.

The terminology of early maps

A knowledge of the technical terms used in connection with early maps is a helpful step towards fuller understanding. These fall into various categories, the first of which refers to the 'state' of the map.

An **impression** is a single example printed from a block or printing plate. Sometimes different impressions, on examination, reveal alterations to the plate; these are said to be in different **states,** i.e. they are made from different states or conditions of the same block or plate. The total number of impressions taken from any plate at one time, together constitutes an **issue,** while further sets of impressions taken from the plate are **reissues.** An **edition** consists of all issues and reissues printed from one state of a block or plate.

Various Latin terms and abbreviations are used on maps to refer to the craftsmen engaged in their production:

(a) The cartographer, or the person responsible for the surveying and for the preparation of the draft map for the engraver, is indicated by one or other of the following: *descripsit, delineavit, delt., del., auctore.*

(b) The engraver is referred to as: *sculpsit, sculp., sc., fecit, caelavit, engr., incidit, incidente.*

(c) The printer or publisher is indicated by: *excudit, excud., exc., sumptibus, ex officina*.

A bound collection of maps is given various titles. *Theatrum* was used by Ortelius in *Theatrum Orbis Terrarum* (1570); John Norden used *Speculum* in the title of his proposed atlas *Speculum Britanniae;* John Speed entitled his atlas *Theatre of the Empire of Great Britaine* (1611). Each of the foregoing terms is used in the sense of a 'view' or 'display'. In 1585 Mercator used the term *Atlas* for the first time and this has remained in common usage ever since.

Other terms occasionally seen are **geographer, cartographer, topographer** or **cosmographer** (maker of maps); **hydrographer** (maker of marine charts); **illumineur** (map colourist).

Identifying an early map

It is not always easy to date a map precisely, for mapmakers often omitted the date of issue and, furthermore, even when the date was included, map plates were often used over and over again without alteration to the date, e.g. a feature may be noticed on a map which could not have existed at the date shown, thus clearly indicating revision of detail since the date shown on the map. One should be aware, then, that the printed date is not necessarily a reliable indication of the date of the topographical information on a map.

Numerous carto-bibliographies are available to assist in identification. The standard work has long been Thomas Chubb's *The Printed Maps in the Atlases of Great Britain and Ireland; a bibliography 1579-1870* but this has recently been superseded in part by R. A. Skelton's definitive work *County Atlases of the British Isles* which only covers the period 1579-1703. On a county level many bibliographies of printed maps are available such as those by Sir H. G. Fordham and Harold Whitaker, or the excellent catalogue of Warwickshire maps by P. D. A. Harvey and Harry Thorpe. When such an aid is not available, C. Verner's 'The identification and designation of variants in the study of early printed maps' (*Imago Mundi* Vol. XIX 1965) provides a useful introduction.

A map can be allocated to a specific period by a study of its style and decoration but more precise dating requires detailed examination of its physical and external characteristics as well as its topographical content. The printing paper should be examined as well as the state of wear on the plate; the imprint; whether text is printed on the reverse and, if so, the language in which it is written. An examination of the watermark in old maps and documents is a well-established method of dating fairly closely, for a particular watermark indicates the age and origin of the printing paper. This is a complex subject, however, and the interested reader is referred to Edward Heawood's paper 'The Use of Watermarks in dating Old Maps and Documents' which can be found in two sources, *The Geographical Journal* Vol. LXIII pp. 391-400, 1924, or in Raymond Lister's *How to Identify Old Maps and Globes*, Bell 1965. In addition to the study of the physical characteristics of the map particular note should be taken of its topographical detail—roads, canals and railways and the extent of any settlement shown.

Map design

(a) The cartouche

Certain components of the design of a map do not form part of the information being presented, but are ancillary to it. On early maps the most prominent of these is the **cartouche,** a panel, often elaborately ornamented, which serves to contain title, key or dedication and which first appeared on maps of the Italian school. On early woodcuts its design was necessarily simple for this technique was unsuited to intricate ornamentation. Copper engraving allowed the craftsmen to develop the ornament of the cartouche, often in a characteristic strapwork design resembling lengths of interwoven leather with curling ends (Plate 2). Such designs are seen on the maps of Saxton and Speed. Another popular cartouche design in the 16th century resembled a carved wooden framework (Plate 3a) with curling projections which supported the title panel. In the late 16th and early 17th centuries, when the Dutch held supremacy in cartography, cartouche designs showed the influence of pattern books of Renaissance sculpture, wood carving, jewellery, stone and plaster work. Designs were influenced too by hammerbeam roofing in Gothic houses and churches with additional embellishments of fish, fruit and flowers. After c. 1580 smaller, equally ornate cartouches were introduced to house the linear scale or the dedication (Plate 5d). The early 17th century saw the disappearance of the pseudo wooden frame and the reintroduction of strapwork, though now in a form reminiscent of plasterwork rather than leather. In the middle of the century, scenes of local life were common and may be seen to advantage on many maps by the Blaeu family.

The Baroque style exerted a strong influence on map decoration with cartouches incorporating masses of flowers and fruit, human figures, animals and architectural detail. The wood carvings of Grinling Gibbons and painted ceilings by Italian artists such as Verrio also played their part in influencing the decoration of maps.

Rococo superseded Baroque in the mid 18th century and is splendidly seen on the maps of Emanuel Bowen. The style was light and elegant and map ornamentation was derived from the drawing room with cartouches which resembled Rococo mirrors or Chippendale chair backs. Influences came also from the countryside—title panels, for instance, often consisted of stone slabs around which country folk displayed their farm implements, produce and animals.

The Romantic movement in England left its mark on map design in the portrayal of classical ruins—pillars framing a title against a background landscape. Many of the one-inch to one mile county maps published during the 18th century had superb cartouches and vignetted scenes of landscapes or buildings. The two fine series of county maps by John Cary and Charles Smith (Plate 15), however, eschewed decoration, their titles being contained within the simplest panels, while the *Old Series* maps of the Ordnance Survey had no decoration whatsoever.

(b) *The border or frame*

Irregularly-shaped maps have been traditionally held together by a rectangular frame, a convention which left unwanted space between the detail of the map and the frame in which the engraver could use his imagination to decorate in one way or another. Ortelius, in the 16th century, filled in the corners of some of his maps with elegant designs which appear to have been derived from metalwork (Plate 5b). Bowen, in the 18th century, filled in spaces with long descriptive notes (Plate 14).

Italian maps of the 16th century had plain borders, sometimes graduated in degrees of latitude and longitude, a device still used today. Maps from the Low Countries had patterned borders resembling moulded picture frames. Ortelius, Blaeu and Visscher among others incorporated local scenes, figures and town plans into some of their border designs (Plate 7a). Another attractive and colourful idea was to introduce heraldic shields into the border—either in regular fashion as in Speed's Cambridgeshire, or interlaced as in Blaeu's map of Piedmont. The practice of using heraldic shields reached its zenith in the 18th century when John Harris included one hundred and fifty-two shields in the border of his map of Kent (1719).

The borders of the early Ordnance Survey one-inch to one mile maps were faintly reminiscent of the piano keyboard, an idea which seems to have influenced county mapmakers such as C. & J. Greenwood, who used variations on this theme.

(c) *The linear scale*

A linear scale plays a functional role on any map but, in the past, it has also served as a decorative feature. Through the centuries it has been common practice to incorporate a pair of dividers into the scale design, sometimes held by small naked figures holding surveying chains and instruments (Plate 5d). Prior to the legal establishment of the statute mile of 1760 yards in 1593, various local or 'customary' miles were in use. Indeed they remained in use for over a century later—on Robert Morden's maps, for example, scales representing 'Great', 'Middle' and 'Small' miles are to be seen, while on some European maps national scales such as 'German', and 'Italian' are included (Fig. 1, page 2). The ultimate confusion to the traveller comes with Julien's map of France (1751) on which appear no less than twenty different scales!

(d) *Orientation*

Maps are normally printed so that north is at the top, but there is really no logical reason why this should be so. Early medieval mapmakers set east to the top in deference to the holy places of the Orient—hence the term *orientation* and the modern sport of orienteering. The practice of orienting maps to the north was established by Italian and Catalan chartmakers but even in the late 18th and early 19th centuries maps were occasionally oriented differently, often to enable them to fit the limited area of the printed page—in Cary's *Traveller's Companion,* for example, the Cheshire map has west to the top.

(e) *Colouring*

Early maps were invariably printed in black and white. Colour printing only came into use in the late 19th century but, traditionally, maps had been hand-coloured at an extra charge and map colouring was a specialised and respected profession. Ortelius, in his early career, was a map colourist and the profession was so highly esteemed in 17th century France that Nicholas Berey was granted the title 'enlumineur de la reine'.

In the 16th and 17th centuries brilliant colouring was lavished on the decorative parts of maps while topographical detail received somewhat lighter treatment. Certain conventions were established—hills brown or green, woodland green, rivers and seas blue, settlements red. Boundaries were normally coloured in outline with a band of colour but the 18th century Germans, Homann and Seutter, brushed a wash over whole provinces, a practice which gave their maps a dull and heavy appearance. Today it is difficult, even for an expert, to establish just when a map has been coloured, for the profession of map colourist still exists, and genuine early maps often receive modern hand colouring. The collector should note that when he sees the term 'contemporary' applied to map colouring it means contemporary with the printing of the map and not present-day. A modern litho-printed reproduction can easily be distinguished from a hand-coloured original if each is examined through a magnifying glass. The hand-colouring will be seen as solid, while the colour printing appears as multitudinous tiny dots due to the use of dot screens in printing.

Several early accounts of map colouring methods survive, the most detailed being in John Smith's *The Art of Painting in Oyl* (1701). A lengthy extract from this work dealing with the colouring of maps appears in Raymond Lister's *How to Identify Old Maps and Globes*, Bell (1965).

(f) *Calligraphy*

The primary requirements of lettering for maps are legibility, perceptibility and suitability, so that in turn it can be clearly read with the naked eye, stand out well from its background, and suit the process by which the map is printed. The early wood engravers favoured the Germanic **Gothic** or **black-letter**, but with the greater flexibility allowed by copperplate engraving, Roman and smoothly-flowing italic alphabets became common in the 16th century. On Flemish and Dutch maps flamboyant **swash** lettering with sweeping tails and flourishes was utilised to fill up unwanted spaces (Fig. 2, page 3, and Plates 1, 2).

Topographical detail
(a) *Seas and coasts*

An essential element of map design is the clear differentiation between sea and land. Colour offers a simple solution but engravers have tried various techniques in an attempt to overcome the difficulty in black and white only. In woodcut maps, exemplified by those of Sebastian Münster, simple wave patterns were engraved, but in the 16th century a stipple pattern of fine dots gave a distinguishing tone to sea areas. This was occasionally implemented by short, horizontal lines drawn outwards from the coast (Fig. 3a). Several maps in Ortelius's *Theatrum* employ more refined wave patterns and perhaps the most attractive device of all was the development

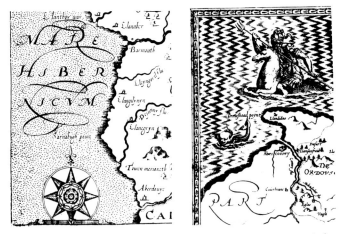

3. (a) *Detail from the map of Merionethshire, engraved by William Kip for Camden's 'Britannia', 1607, illustrating the use of stipple on sea areas (an easy technique for the engraver), combined with short, horizontal lining to accentuate the coastline. Hills are crudely engraved with shading on both sides.* (b) *Detail from Speed's Denbighshire map, 1611. The engraver was Jodocus Hondius and the type of sea shading shown here, commonly used by him, is reminiscent of moiré or watered silk.*

by Hondius of the striking design, resembling moiré or watered silk (Fig. 3b).

During the late 16th and early 17th centuries it was common practice to fill the seas with monsters or beautifully engraved sailing vessels, though Blaeu and his rival, Jansson,

4. *Delicately engraved sailing vessel from a map by Blaeu, 1645.*

often left seas blank except for the occasional ship and horizontal lining round the coasts. The depiction of sailing craft of various kinds on early maps would, incidentally, make an interesting study in itself.

Form lines, i.e. lines drawn parallel with the shore and gradually increasing in distance apart, were used to accentuate coastlines in the 18th century and this practice reached a peak of perfection in early Ordnance Survey maps.

11

(b) Mountains and hills

The depiction of relief has always posed problems for the mapmaker in that, ideally, he has to show heights accurately as well as length and breadth. On medieval manuscript maps hills were crudely drawn in profile with shading on the sides to add substance (Fig. 3a). Early engraved maps, e.g. Saxton and Speed, had hills like sugar-loaves, shaded to one side, with no attempt at vertical scale but some effort at relative heights (Fig. 8). Blaeu and his contemporaries were no nearer to a solution in the mid 17th century, merely drawing smaller hills and more of them. Pictorial representation continued towards the end of the 17th century when **hillshading,** i.e. lines drawn down the slopes, was employed in an effort to show hills in plan. Steepness could be suggested by thickening the lines but height could still not be demonstrated. In the 18th century ranges of hills appeared as 'hairy caterpillars' but, at the turn of the century, the technique of **hachuring** (a more scientific development of hill-shading) came into being, sometimes in conjunction with **spot heights.** The **contour** method, by which height or depth could be indicated effectively, was first used by a Dutchman, N. S. Cruquius, to chart a river bed in 1729, and in 1737 Philippe Buache depicted the depths of the English Channel by contours. On land maps, however, contours did not come into use until well into the 19th century.

(c) Woodland and parks

The conventional representation of woodland has changed very little as it was customary from early times to draw trees stylistically in elevation and to colour them green. Saxton introduced a symbol for great parks in the form of a group of trees surrounded by a pale fence.

(d) Settlement

Towns received pictorial treatment in elevation on medieval manuscript world maps but since then there has been a gradual change, first to the bird's eye perspective view and later to depiction completely in plan. The Bodleian map of the British Isles (c. 1360) uses pictorial symbols for varying categories of town—ordinary towns have cream-washed buildings with red tiled roofs; small monastic towns have buildings with spires; cathedral towns red and white roofed buildings with soaring spires; while London receives individual treatment, its buildings having blue leaded roofs, painted timber beams, spires, crosses and battlements.

During the 16th century mapmakers tried to show towns

partly in plan and partly in elevation by raising their viewpoint and drawing in bird's eye perspective. This treatment, which can be very effective, is well seen in some of the plans in Braun and Hogenberg's *Civitates Orbis Terrarum* (1573). It is particularly good at demonstrating the character and architecture of a town. On small-scale maps pictorial symbols made up of a group of buildings showed whether a town had fortifications or not and indicated its relative size by the number of buildings in the group. Following the decline in importance of the medieval castle, it was superseded by the church as a symbol for settlement. A map particularly remarkable for its pictorial treatment of village churches is Philip Symonson's superb county map of Kent (1596). The delicate engraving of town symbols was often obscured by the tendency to colour them too thickly in red but, nevertheless, this practice continued until the late 18th century.

5. (a) *Table of settlement symbols used on a map of Osnaburg by Ortelius, 1573. (b) Table of symbols used by William Smith on his map of Northamptonshire, 1602. The use of a small circle in conjunction with engravings of buildings to indicate types of settlement is clearly shown.*

From early times the circle has been associated with settlement on maps and from 1520 it was used in conjunction with church towers or other buildings to symbolise towns and cities—the circle being placed to indicate the town centre and the point from which distances should be measured.

German mapmakers such as Philipp Apian in the mid 16th century introduced numerous symbols for settlement which

13

6. *Town plan of Denbigh from the map of Denbighshire by John Speed, 1611. The town is portrayed in bird's eye perspective and principal streets and buildings are given an identifying letter which is explained in the key.*

were explained in a table of conventional signs or **legend.** Similar tables were introduced into England by William Smith (Fig. 5b) and John Norden.

John Speed included inset plans of county towns on the maps in his *Theatre of the Empire of Great Britain* (1611) and, in so doing, produced the first extensive set of printed British town plans (Fig. 6). As we have already noticed, marginal town views were introduced into the maps of Ortelius, Blaeu, Visscher and other mapmakers. These presented useful additional information to the detail included on the maps themselves (Plate 7a).

The period from 1675 to 1740 was one of transition when scientific plans such as that of London by Ogilby and Morgan appeared at the same time as many traditional bird's eye views and elevations. On small-scale maps the larger towns were drawn in plan by means of blocks of buildings with white roads threading their way through; smaller towns and villages were symbolised by rows of houses drawn in perspective along the road sides.

After 1740 the practice of drawing settlements in plan became much more common and with the advent of the Ordnance Survey one-inch to one mile maps in the early 19th century it became the standard practice.

THE ORIGINS AND EARLY DEVELOPMENT OF MAPMAKING

How and where maps were first made is something about which we can only surmise. We know, however, that primitive peoples have an inborn ability to make rudimentary maps of their surroundings and this has led to the assumption that the making of simple sketch maps is older than writing and that maps are, therefore, older than written history. The earliest surviving map is inscribed on a small clay tablet found at Gar-Sur, two hundred miles north of Babylon. It dates from the third millenium B.C. and shows a river, probably the Euphrates, flowing out through a delta and flanked by mountain ranges resembling fish scales. The Babylonians achieved the first essential of mapmaking, that of establishing the position of any place so that it could be located at any time, for they divided the circle of the sky into 360° and the day into hours, minutes and seconds, thus enabling any point on the earth's surface to be plotted in relation to the stars.

The ancient Egyptians measured and recorded land for taxation purposes but we owe the true foundations of scientific mapmaking to the Greeks. Eratosthenes, accepting the theory that the world was round, calculated a remarkably accurate figure for its circumference but a later astronomer, Posidonius, made new calculations which were seven thousand miles too low. These later figures were accepted and exerted an unfortunate influence on maps for centuries. The greatest Greek geographer was Claudius Ptolemy (AD 90 to 168) whose famous work *Geographia* discussed the construction of globes and map projections and provided a list of eight thousand places with their latitudes and longitudes. The world could now be fitted into a scientific framework to which new discoveries could be added as they were made.

There were, however, errors in Ptolemy's figures—because he used the erroneous calculations of Posidonius, he underestimated the earth's size. He believed that Asia and Europe occupied more than half the circumference of the globe, and he calculated the length of the Mediterranean to be 62° instead of its correct 42°, an error which persisted on maps derived from his data until 1700. *Geographia* includes twenty-six regional maps and a world map in what is known as the A recension, and a group of sixty-seven maps of smaller areas forms the B recension. As there are, however, no surviving manuscript maps based on Ptolemy earlier than the 12th century, it is not known whether Ptolemy drew maps himself or whether they have been merely ascribed to him on his reputation as a geographer. *Geographia* was lost to the western world for centuries, though it exerted an influence on Islamic geography, and was only brought back to Europe in the 15th century when it played a major role in furthering the great renaissance of cartography.

Medieval cartography

The Romans made surprisingly little contribution to cartographic progress, but after the fall of the Roman Empire the spread of Christianity led to the appearance of new maps which were centred on Jerusalem. Monastic writings were often illustrated with maps known as T-O or T in O maps which were circular in shape with east to the top and Jerusalem at the centre. The ocean flowed round the circumference to form the O while the T, placed inside the O, divided the map into three parts. The vertical stroke of the T was formed by the Mediterranean Sea and the horizontal bar consisted of a line from the Don to the Nile (Fig. 7). Such medieval maps were produced in great numbers, often simple in form but, in some cases, with a wealth of detail, elaborate decoration and brilliant colouring.

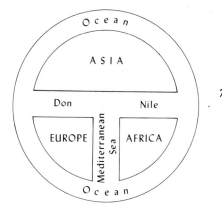

7. *Diagram to illustrate the principle of a T-O or T in O map.*

Matthew Paris

Among the finest maps of the early Middle Ages were the four of Great Britain made by Matthew Paris, a monk of St. Albans, c. 1250. Although they were crude in detail the maps present a readily recognisable portrait of the country and are constructed around the pilgrim route from the north to Dover. The stations along this route are laid down along the central vertical axis and, to conform to this axial route, Matthew had to resort to some distortion—the estuary of the Thames, for instance, is on the south coast. Nevertheless, Matthew obviously recognised a fundamental principle of mapmaking—that of drawing to a uniform scale—and, furthermore, he included a lot of detail, naming over two hundred and fifty places as well as rivers and a few hills.

The Hereford and Ebstorf maps

One of the finest medieval world maps is preserved in Hereford Cathedral and was made c. 1300 by Richard of Haldingham. Measuring over five feet in diameter, this colourful circular map illustrates the Biblical world, centred on Jerusalem, with the figure of Christ presiding at the top (or east). It is richly decorated with detail from Biblical lore, medieval histories and bestiaries. Representation of the British Isles is very poor, for they appear to the edge of the map and have been bent round to fit into the circle, with a consequent distorted shape on which, nevertheless, some new place names are included. The Ebstorf map of the same period was even larger and more magnificent, measuring over thirteen feet in diameter. It was built around the figure of Christ, on a fairly standard T-O pattern with a richly gilded Jerusalem in the centre. The map was formerly preserved in a monastery at Ebstorf in Germany but was regrettably destroyed during the second world war.

The Gough or Bodleian map

First described by Richard Gough in 1780 and now in the Bodleian Library, this fine map was made c. 1360 by an anonymous mapmaker and even a casual glance reveals the tremendous advance it represents. A map of the British Isles, 3ft. 9½in. by 1ft. 10in., it was made on two joined skins of vellum and, though the north-south extent is exaggerated, the shape is remarkably good. Its most important cartographic innovation is the accurate depiction of roads and distances.

The Portolan charts

Although contemporary with medieval *mappae mundi* the portolan charts represent a different tradition, surpassing in accuracy anything which had gone before. At first the work of Italian and Catalan cartographers (later also of Portuguese) the charts were naturally concerned with the faithful delineation of coasts and harbours, land areas being usually almost blank. Most of these marine charts were made on sheepskin and covered a similar area, the Mediterranean and Black Seas well depicted and the Atlantic as far north as Ireland not so satisfactorily. With new geographical discoveries the area shown was gradually extended, the first example to show the British coastline being the *Carte Pisane* (c. 1300). The most striking feature of the charts is the system of lines of compass directions (**rhumb lines**) and compass **roses which covers their surface and was presumably** intended to assist the mariner in setting his course.

The Catalan Atlas

The peak of the portolan tradition was reached in the Catalan Atlas (c. 1375) made, it is believed, by a family of Catalonian Jews working in Majorca, for Charles V of France. The Catalan Atlas is a world map of twelve leaves and abandons the circular form of the earlier *mappae mundi*. For the first time Asia is seen in recognisable form, its detail based to some extent on Ptolemy but with additional material obtained from the journeyings of Marco Polo.

The re-discovery of Ptolemy's Geographia

The period around 1500 was one of great significance to cartographic development and three events in particular were to play their part in furthering a renaissance of cartography. First was the re-introduction of Ptolemy's *Geographia* into Europe which meant that the religious attitude to mapmaking was replaced by scientific principles. The second stimulus came from the great discoveries, for new information brought by Vasco da Gama, Cabot, Columbus and others meant that geographical knowledge increased at an unprecedented rate. This was reflected in improved world maps by Juan de la Cosa (1500), Martin Waldseemüller (1507) and Diego Ribero (1529). Most significant of all, however, was the third event —the invention of printing. This revolutionised mapmaking more than any other single event at any time for it made maps available to an enormously wide readership. No longer was it necessary to copy maps laboriously by hand—instead, multiple copies could be taken, at first from wood blocks and later from copper plates.

THE PRINTED MAP

The early German school

Like other branches of art, mapmaking has tended to fall into national schools. In the late 15th and early 16th centuries, the centres of geographical learning were in the southern German cities of Augsburg and Nuremburg. Here, also, the wood-engraving craft flourished with famed artists such as Dürer and Holbein at work. In Nuremburg in 1493 Hartmann Schedel produced his famed *Nuremburg Chronicle* with maps and town views engraved by Dürer's master, Michael Wohlgemut. Nuremburg was famous for fine globes, including that by Martin Behaim (1492), and here also were produced primitive road maps. Woodcut editions of Ptolemy were published at Ulm in 1482 and at Strasbourg in 1513, the latter including an early attempt at three-colour printing. In 1528 Sebastian Münster planned a composite atlas under his own editorship and invited German geographers to send him maps of their own provinces. Much of the resulting material appeared in Münster's edition of Ptolemy (1540) and his *Cosmographia* (1544).

The Italian school

To Italy goes the credit for the revival of interest in Ptolemy's classic *Geographia* and early editions with copper-plate maps were printed in Bologna (1477), Rome (1478), Florence (1482) and Rome (1490). These were fine examples of Italian craftsmanship. In the early 16th century the Italian Renaissance was at its peak—Italian craftsmen were supreme and so, too, were Italian mapmakers. The map trade flourished in Rome and Venice where men such as Lafreri, Bertelli and Camocio may well be termed 'publishers' for they combined cartography, engraving, printing, publishing and mapselling. Perhaps the finest of the Italian cartographers, however, was Giacomo Gastaldi whose prolific output included the maps for the 1548 edition of Ptolemy.

The Italian publishers may be credited with producing the prototype of a modern atlas for they assembled in one volume, to order, collections of maps from their stock. Sixty to seventy such atlases survive, with maps arranged following the order of Ptolemy, and these have been termed 'Lafreri Atlases'.

The Netherlands

With Holland becoming a dominant commercial and naval

power in the late 16th century, the Dutch assumed cartographic supremacy from the Italians. There began what has been called a golden age of cartography when the design and decoration of maps attained unprecedented heights. Gerhard Kramer, known to us as Mercator, is usually known as the father of Dutch cartography. He was a man of great talent; land surveyor, instrument maker, engraver, cartographer and scribe. In addition to the projection which bears his name he was responsible for some major cartographic improvements including the correct measurement of the length of the Mediterranean. His maps are notable for their calligraphy and it is typical of the man that, being unsatisfied with any existing alphabet, he devised a new one for himself and published a treatise about it in 1540. His collection of maps *Atlas sive cosmographicae meditationes de fabrica mundi et fabricata figura* (1585-95) is particularly significant because this was the first time that the term 'atlas' was used in relation to maps.

Mercator was followed by a succession of illustrious cartographers. The first was Abraham Ortelius whose *Theatrum Orbis Terrarum* (1570) consisted of seventy maps gathered together from the best available sources. Jodocus Hondius was another renowned Dutch mapmaker who took over Mercator's business after the latter's death. An engraver of unrivalled quality, Hondius was responsible for engraving the plates for John Speed's fine series of county maps. In Amsterdam at the middle of the 17th century the rival firms of Blaeu and Jansson were producing atlases of incomparable beauty and elegance (Plates 6, 7, 8). The Visscher family too published fine work during the 17th and early 18th centuries and other great names in Dutch cartography were Plancius, Linschoten, De Jode, Waghenaer (who produced the first great printed sea atlas *Spieghel der Zeevaerdt* at Leyden in 1584-85), Pieter Goos and Frederick de Wit.

The French school

France had no outstanding record of cartographic achievement prior to the 17th century but for two centuries afterwards a school of mapmaking flourished in the country, its leading light being Nicholas Sanson (1600-1667) who was succeeded by his sons, Adrien and Guillaume, and thence by various descendants, to continue an important cartographic dynasty which produced a profusion of fine maps and atlases. Sanson was no lone figure: there were numerous other fine

French mapmakers such as Alexis Hubert Jaillot (1640-1712), whose maps were regarded by Sir George Fordham as being among the finest specimens of decorative cartography.

England

The first copper-engraved map of the British Isles to be based on contemporary information appeared in 1546 and is attributed to George Lily, a Catholic refugee living in Rome. In the late 16th century, Laurence Nowell, Dean of Lichfield, prepared a manuscript map of England, Wales and Ireland which was the most accurate of its time and also had an ambitious scheme for a survey of all the English counties which he proposed to Sir William Cecil. The completion of such a scheme, however, had to await the surveys of the Yorkshireman, Christopher Saxton.

BRITISH REGIONAL MAPPING: 1579-1700

Before the late 16th century Britain had been mapped as a single unit rather than on a regional basis, but from Elizabethan times until the beginning of the 19th century the county became the basic unit for mapwork and it was only the advent of the Ordnance Survey with its one-inch to one mile map on national sheet lines which marked the beginning of the end for the county map and the private county surveyor. By 1570 the time was ripe for cartographic development in this country. Over in Europe Flemish mapmakers, with their skill in the newly-discovered art of copper engraving, had wrested cartographic supremacy from Rome and Venice, and a number of Flemish Protestant refugees began to introduce these new talents into England. Surveying methods and instruments were improving rapidly and various treatises on the art of surveying were published such as the *Pantometria* (1571) of Leonard Digges in which he describes an instrument of his own invention resembling a modern theodolite. The Elizabethan age was one of commercial prosperity and also one of intellectual attainment. The arts were sponsored by landed gentry who came into estates as a result of the confiscation and redistribution of monastic lands. This reallocation of land also led to a demand for new estate maps, a demand which brought about the creation of a new body of professional surveyors, known as 'land-meaters'. From the ranks of this body emerged one of the greatest figures in British cartography—Christopher Saxton.

Saxton's 'Atlas of England and Wales' (1579)

Saxton, born in the hamlet of Dunningley near Wakefield c. 1542, was commissioned by a Suffolk gentleman, Thomas Seckford, to survey and map all the counties of England and Wales. Saxton accomplished this daunting task at a remarkable rate with the result that his first two maps were engraved in 1574, followed by others in rapid succession so that the complete work, consisting of a general map of England and Wales and thirty-four county maps, was published in 1579 as *An Atlas of England and Wales*. To Saxton, and of course to Seckford, without whose patronage the work would have been impossible, goes credit for producing what was the first national atlas to be produced by any country. Yorkshiremen can be proud of Saxton and he, in turn, seems to have been inordinately proud of his birthplace for the name of Dunningley, representing only a cluster of houses, appears not only on his fine double-page map of Yorkshire but also, more surprisingly, on his general map of England and Wales. A ten-year privilege was granted to Saxton by Queen Elizabeth to publish and market his maps and he was also given his own coat of arms, the sole English mapmaker to receive such an honour. The Queen had shown interest in the project throughout and Saxton included the Royal Tudor arms on each map together with those of Seckford.

Not a great deal is known about Saxton himself or his methods of work but it is clear that he was an original surveyor, travelling the countryside on horseback and making compass sketches or using a plane table to plot rough maps. To assist him, an open letter was sent by the Queen to local mayors and justices of the peace ordering that Saxton should be conducted to any suitable high place from which he could view the country and that he should be accompanied and assisted by the most knowledgeable local people. Notwithstanding this assistance there is little doubt that Saxton must have made reference to earlier topographical works such as the manuscript itinerary of John Leland and the *Perambulation of Kent* by William Lambarde. Without such references Saxton could hardly have completed his survey so quickly.

It was not easy to find English engravers skilled enough in the newly-found craft of copper engraving to engrave the draft maps surveyed by Saxton and to overcome this problem, Seckford turned to Flemish Protestant refugees. Nine maps in the atlas were signed by Remigius Hogenberg, five by Leonard Terwoort—a particularly flamboyant engraver (Plate 4a) and two by Cornelis de Hooghe. Due to their work on this atlas the skill of the Flemish craftsmen became fully recognised in this country. Three English engravers did work for Saxton—Augustine Ryther, whose signature 'Anglus' distinguished him from the foreigners, Francis Scatter and Nicholas Reynolds. Their maps were more restrained in their decoration than those of the Flemings but Ryther's maps, in particular, are among the finest in the atlas. In general the Saxton maps are among the most decorative of all time, the product of an exuberant age, displaying a wealth of ornamentation to which brilliant hand-colouring was often added after printing.

The maps have a freer style of lettering than that used on woodcut maps for the graver used in copper engraving could be handled almost as easily as a pen. Nevertheless it should be borne in mind that the craftsman, when engraving the thick copper sheets, had perforce always to work in reverse so that the printed map subsequently made from the copper engraving had all its detail reading the right way round. The restraint of the fine script on the face of the maps is offset by the flamboyant swash lettering used to fill up sea areas and to indicate the names of adjacent counties (Plate 1).

Saxton made good use of conventional symbols to represent the human and natural features of the landscape but provided no explanation or legend. A prominent feature is the use of conical sugar-loaf hills, shaded on the eastern slopes and coloured green or brown (though the colouring has no particular significance). There is no attempt at scale in the hill depiction though there is some evidence of an effort to indicate relative size—e.g. Pendle Hill and Ingleborough, on the Lancashire map, appear taller than the rest. Rivers are clearly shown by a double line with bridges indicated by two short parallel lines across the course of the stream. Roads are surprisingly omitted, perhaps implying that Saxton intended his work more as a record of the relative position of places and a depiction of the landscape than as a practical aid to the traveller. Woodland is symbolised by groups of delicately engraved trees and, if maps by different engravers are studied, it will be seen that each used slightly varying versions of the tree symbol. Great parks are graphically represented by a pale fence surrounding a cluster of trees.

Saxton obviously regarded the positioning of settlements accurately as of prime importance and used a dot within a small circle to show the centre of each place. This circle forms part of a more elaborate symbolisation for villages and towns; villages being shown by a church or tower in elevation; towns by a small group of buildings and cities by a sizeable cluster of buildings and churches.

County boundaries are prominently drawn and often emphasised by a band of colour. Hundred boundaries, on the other hand, are shown on only five maps, despite the fact that the hundred remained an important administrative unit at the time.

Sea areas presented a challenge to the engravers' flair for extravagant display and, though the sea itself was shown soberly enough by a fine dot pattern, a variety of craft sailed across its surface together with frolicking mermaids, sea monsters and fish. It was often in the sea areas, too, that the engraver placed elaborate cartouches, heraldry, mile scales and similar features.

The words *Christophorus Saxton descripsit* appear on each map together with Seckford's motto *Pestis patriae pigricies* on the early maps, the latter being altered on later maps to *Industria naturam ornat*. The maps are enclosed in narrow borders resembling moulded picture frames in a variety of styles. Each map measures $18\frac{1}{4}$ in. by 15 in. and all were designed to fit the same page size in the atlas. Consequently there are variations in scale from approximately four miles to one inch to two miles to one inch. Some of the smaller counties are grouped two, three, four, or even five, to a page.

A complete Saxton atlas, in the unlikely event of one appearing for sale, would command a very high price indeed today and even individual maps are expensive. The price of the latter varies with what the demand is likely to be—a populous county such as Essex is certain to be priced much higher than a county with a small population like Merionethshire.

Philip Symonson's map of Kent, 1596

Although Symonson did not produce a series of county maps, his map of Kent, published in 1596 and engraved by Charles Whitwell, is worthy of special mention for it is regarded as the finest example of Elizabethan mapmaking. It was printed on two large sheets but only the eastern half of the county has survived. The map is exceptionally detailed and, like the maps of Symonson's contemporary John Norden, shows main roads. Churches are carefully and delightfully delineated, as are the windmills of the county. A further interesting innovation is the distinguishing of the navigable portion of the river Medway.

John Norden's 'Speculum Britanniae'

John Norden (1548-1626) was a Somerset man and, like Saxton, was an estate surveyor of considerable ability as well as being a cartographic innovator. Unlike Saxton, however,

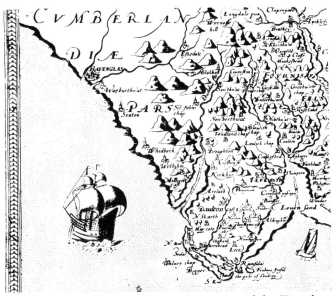

8. *Section of the map of Lancashire engraved by Hogenberg for Saxton's 'Atlas of England and Wales', 1579. Saxton's attempt to give some indication of relative heights can be well seen.*

he was plagued by financial difficulties and his scheme for a series of county histories with maps, entitled *Speculum Britanniae*, failed through lack of patronage when only five counties had been completed. Norden, conscious of the deficiencies of other atlases—the lack of roads on Saxton's maps; the need for an English translation of Camden's *Britannia*, and the weight and bulk of these tomes—wished to produce a work which would be more useful to the general public, providing improved maps together with information about the archaeology, industry, agriculture and history of each county. Only two sections of *Speculum Britanniae*, those dealing with Middlesex and Hertfordshire, were published in Norden's lifetime. His survey of Cornwall was not published until 1728, Essex not until 1840. Norden also drew larger maps, those of Sussex, Surrey and Hampshire being printed. His maps were the first county maps to show roads though Emmison and Skelton state that their examination of four copies of the Essex map reveals so many discrepancies in the mapping of Essex roads that their portrayal must be treated with caution[1]. The triangular table of distances, by which distances between places could be easily read off, was introduced into this country by Norden and is still effectively used today. So too is his grid reference system, dividing the maps into squares in which the vertical lines are allocated a number and the spaces a letter so that any place can be quickly located by reference to its grid square. Tables of conventional signs were included, listing such items as 'market townes, parrishes, hamletes, noblemen's howses, howses of gent', etc.', and the like. The quality of Norden's work is evident from the fact that later mapmakers copied his maps, when available, in preference to those of Saxton.

William Smith's maps of 1602-3
(formerly known as the Anonymous Series)

These maps of twelve counties, apparently designed to form part of a single series, for long remained the work of an unknown mapmaker, but are now credited to the herald, William Smith (1550-1618) who, apart from this series, produced a *Description of England* containing plans and profiles of towns. Smith's handsome county maps are based on Saxton and Norden's work but incorporate additional place names, roads, hundred boundaries and names, and a table of conventional signs. They are engraved in the Flemish style with outstandingly good calligraphy and are probably the work of Jodocus Hondius. Compared with Saxton and Speed's maps, their ornamentation is restrained.

William Camden's 'Britannia' 1607

The 1607 edition of *Britannia* by the antiquary, William Camden, included a set of county maps which were notable in that, unlike the Saxton atlas, each county was given an individual map. The Camden maps are based on Saxton and Norden and the unwary collector is sometimes misled into thinking that they are original Saxtons, a trap that is easily avoided for the two sets are markedly dissimilar. Camden's are much smaller, measuring only c. 10in. by 14in., and were engraved by William Kip and William Hole. They are much less detailed and not so decorative as Saxton's maps. Nevertheless they are worth collecting for the clarity of their engraving and their attractive appearance, particularly when hand-coloured (Fig. 3a).

John Speed (1552-1629)

From the early 17th century, mapmaking in Britain came

'The Description of Essex' by John Norden, F. G. Emmison and R. A. Skelton. *The Geographical Journal*, Vol. CXIII, Part I, March 1957.

under the dominance of London publishers, and it was John Sudbury and George Humble who launched the most popular British cartographic venture of all time, John Speed's *The Theatre of the Empire of Great Britaine*.

Speed, born at Farndon in Cheshire, was a tailor's son and for a time followed his father's profession, being admitted to the freedom of the Merchant Tailor's Company in 1580. After his marriage Speed lived in Moorfields and his great enthusiasms were drawing maps, writing religious works and studying antiquity. His antiquarian pursuits found him a patron in the person of Sir Fulke Greville who provided a stipend so that Speed might devote his time to historical research and mapmaking. Speed's *Theatre* was intended to illustrate his *History of Great Britaine* but the latter is now forgotten while the maps grow ever more popular. Speed was not an originator, as he freely admits in his oft-quoted remark 'I have put my sickle into other mens corne'. Unlike Saxton, the practical surveyor travelling around, making surveys and accumulating information, Speed was a compiler who worked in library and study assembling his material and preparing rough layouts for the engraver showing all the information to be mapped. These would be engraved in reverse on the copper plate after which the final printed sheets would be run off.

Speed's maps were highly decorative, being engraved in Amsterdam by the renowned craftsman, artist and scholar, Jodocus Hondius, whose skill in blending the material supplied by Speed into a coherent and attractive design entitles him to a sizeable share of credit for the maps' success (Plate 2).

Speed's principle contribution to cartographic development was the inclusion of a plan of the county town on each map. These plans were culled from cartographers such as John Norden, William Smith and William Cunningham and were engraved in bird's eye view (Fig. 6). They form the first comprehensive collection of English and Welsh town plans and were a valuable step forward in a stagnant era of British mapmaking. Portraits of significant historical figures associated with particular counties were also inserted on the maps and occasionally there are engravings of buildings, such as Old St. Paul's and Westminster Abbey on the map of Middlesex. Hundred boundaries were included but not roads, and the detail generally resembles Saxton's with sugar-loaf hills, woodland and parks, with churches or grouped buildings to indicate settlement. The hills emphasised by Saxton remain prominent but the number of smaller hills is reduced, to little detriment for the hummocky symbols used by Saxton were merely an indication of hilly country and not a precise portrayal of individual hills. Speed's maps were re-issued and re-printed many times until the mid 18th century. They were, of course, issued in black and white only with colouring applied by hand to order.

Michael Drayton's Poly-Olbion maps, 1612-22

Poly-Olbion was illustrated by a set of eighteen curious little regional maps, engraved by William Hole, which have little cartographical interest but are quite the oddest set of county maps ever to be issued (Plate 3b). Topographical detail consists only of rivers, woods, hills and a few towns. No boundaries are shown and each map covers parts of several counties. Allegorical figures, water nymphs, hunters and animals are posed over the landscape and the maps have neither title nor scale.

Pieter van den Keere—the miniature Speed atlas of 1627

Reduced copies of Saxton's county maps were engraved in 1599 by Pieter van den Keere, or Petrus Kaerius, for a proposed pocket edition of the atlas. This came to naught, however, and the plates came into the hands of Speed's publisher, George Humble, who in 1627 issued forty maps by van den Keere, together with twenty-three new maps copied from Speed, as a pocket atlas for the general public—*England Wales Scotland and Ireland described and abridged . . . from a farr larger volume done by John Speed*. The maps are attractive with plain cartouches but their small dimensions, $4\frac{1}{4}$in. by 3in., mean that some counties look very crowded with place names (Plate 3a). Each map was accompanied by descriptive text from Speed's larger atlas.

Thomas Jenner

In 1643 Thomas Jenner, a London bookseller, printseller and engraver, issued *A Direction for the English Traviller* consisting of a set of small, thumb-nail maps, each with a triangular distance table, first engraved by Jacob van Langeren in 1635. Jenner also published a notable map of England and Wales, known as the *Quartermaster's Map* because of its use in the Civil War, which was engraved by Wenceslaus Hollar in six sheets, folding in such a way that it would fit handily into a pocket.

The Blaeu family

In the early 17th century the centre of the international map trade moved from Antwerp to Amsterdam. The renowned publishing houses of Blaeu and his rival, Jansson, had their businesses here, each combining mapmaking with general publishing, printing and bookselling.

Willem Janszoon Blaeu was born in 1571 and in his youth acted as assistant to the Danish astronomer, Tycho Brahe. The knowledge he gained from Brahe enabled Blaeu to set up a business in Amsterdam as a maker of mathematical instruments, an engraver and a map printer. He achieved his personal ambition of housing under one roof an establishment where every phase of map production could be carried on efficiently and his printing house was furnished with no less than nine type presses for letterpress printing of text, six presses for copperplate printing, and a type foundry. A contemporary account gives some insight into the organisation of a 17th century cartographic establishment. The printing house fronted on to a canal and in a room overlooking it were stored the copper plates from which maps and atlases were printed. Adjoining was a press room used for plate printing with a room nearby where book printing took place. To the rear was a room for the storage of type and other printing materials, while upstairs a small room was set aside for the proof readers. The letters used in printing various languages were moulded in the type foundry in an upper storey.

After his death in 1638 the elder Blaeu's business passed to his two sons, John and Cornelius, and later to his grandson. In 1645 John (Joan or Joannes) Blaeu published a set of maps of English and Welsh counties as Part IV of *Theatrum Orbis Terrarum, sive Atlas Novus.* These were derived from Speed but set new standards in craftsmanship and design. Scotland and Ireland were not included but appeared in 1654 as Part V of *Atlas Novus* and in 1662 as Vol. VI of the *Atlas Maior* or *Grand Atlas.*

While the Blaeu county maps may be regarded as superb examples of engraving and design, they add little to our topographical knowledge or cartographic development. Blaeu still made no attempt to show roads and he dispensed with Speed's inset town plans. The detail is generally similar to earlier maps but neater and more delicately engraved. Symbols were smaller and there was no extravagant ornament for its own sake, even sea areas being sparingly adorned. The calligraphy is especially beautiful and so, too, is the way in which the several elements of each map are blended into a logical, balanced composition (Plates 7, 8). The *Atlas Maior,* in sumptuous bindings, was in its day a traditional gift to visiting royalty.

Joannes Jansson (1596-1664)

In 1646 the rival house of Jansson issued a very similar atlas to that of Blaeu. The two establishments had competed—not always amicably, for they were ever ready to criticise each other's work—to produce an atlas containing county maps of the British Isles, and Blaeu's was published a year ahead of Jansson's. Like his rival, Jansson based his work on Speed and, if the two sets of maps are compared it will be observed that the Dutch engravers, perhaps not surprisingly, made several errors when transcribing British place names. In Cheshire, for instance, the engravers have in over twenty cases either wrongly transcribed a letter or arbitrarily abbreviated a name. Jansson was more flamboyant in style than Blaeu and his seas were adorned with compass roses and sweeping lettering.

A feature of both the Blaeu and Jansson atlases is the architectural composition of their title pages, similar in conception though different in execution. Each has an architectural design supporting the royal arms and incorporating figures of Danes, Romans, Normans and Saxons in recesses with the figure of Britannus prominent. Each design has a cartouche below with the imprint and a rectangular title panel. Jansson's design is the more richly ornamented and makes a fine composition. The title pages of these great atlases would make a rewarding study; a later example of an architecturally inspired title page design is provided in Jaillot's *Atlas Nouveau* (1689) and it is revealing to observe how these three mapmakers have interpreted a similar theme.

Both Blaeu and Jansson's atlases appeared in numerous editions and in several languages, the text matter being printed on the reverse of the maps.

Richard Blome

Blome, a cartographer and compiler of books on heraldry and topography, issued *Britannia,* containing a set of county maps, in 1673. The maps were of indifferent quality and their popularity with present-day collectors is hard to understand. While they have a superficially quaint appearance which seems attractive, closer examination reveals them to be poor in execution with crude lettering and ornamentation (Plate 9). Like much of Blome's work they were pirated from earlier sources.

John Ogilby (1600-1676)

The first half of the 17th century was a period of stagnation as far as any serious progress in British mapping was

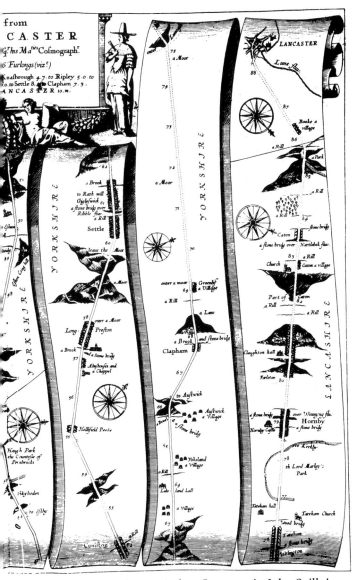

9. *Section of the road from York to Lancaster in John Ogilby's 'Britannia', 1675. Direction is shown by the series of compass indicators. The curious effect of a hill drawn upside down is seen where the road enters a depression, e.g. at Caton on the right hand strip.*

concerned. The map trade was centred on Holland and such English maps as were produced were generally rehashes of Speed, Norden and Saxton. In 1675, however, the publication of *Britannia . . . or an Illustration of the Kingdom of England and Dominion of Wales: By a Geographical and Historical Description of the Principal Roads thereof. Actually Admeasured and Delineated in a Century of Whole-Sheet Copper Sculps . . . by John Ogilby, Esq.* provided a new landmark in the story of English cartography. It was completely new both in conception and execution, the work of a man who was at various times dancing master, theatre owner, translator of Greek and Latin and a deputy master of the revels in Ireland. He experienced the extremes of fortune, first losing his money in the Irish rebellion of 1641 and later facing ruin when the stock of his printing and bookselling business was destroyed in the Great Fire of London in 1666. After the Great Fire he was appointed along with William Morgan, John Oliver and Thomas Mills to survey the devastated areas of the City. The four prepared a detailed plan at twenty-five inches to one mile which was published by Morgan in 1677, a year after Ogilby's death.

By 1670 Ogilby had already begun to contest Dutch supremacy in county mapmaking and he produced fine maps of Kent, Middlesex and Essex in which roads were incorporated. The portrayal of towns on these maps marked the beginning of the change from elevation to plan.

Ogilby was keenly aware of travellers' need for improved delineation of roads and he conceived the revolutionary idea of making up a road book, to be based on original survey, which would present the roads on a series of strips or scrolls. He received the royal approval from Charles II and was given a royal warrant commanding that he be given free access to church books and public records and be informed about significant features in each locality. The surveying of principal roads was a task of great magnitude and Ogilby describes his methods in the Preface to *Britannia*. The road measurements were carried out with a 'wheel dimensurator' or 'perambulator', an instrument similar in principle to a modern cyclometer and consisting of a large wheel, mounted in shafts, which was pushed along by surveying assistants, the revolutions of the wheel automatically registering on a dial the distance covered. A feature of the survey was the use of the statute mile of 1760 yards which had been introduced by Act of Parliament in 1593. Ogilby thus dispensed with the confusion caused by the use of local or 'customary' miles, varying in length from 2035 to 2500 yards. Ogilby's measure-

1. Section of the map of Devonshire from Christopher Saxton's 'Atlas of England and Wales' (1579) showing ornamental border, stippled sea, sea monster and sailing vessels. Roads are omitted but bridging points over the rivers are indicated. The hills of Dartmoor and Exmoor are crudely shown with shading on the eastern slopes. 'Swash' lettering fills in the space between county boundary and frame.

2. *Eastern section of Buckinghamshire from John Speed's 'Theatre of the Empire of Great Britaine' (1611). Hundred boundaries are clearly shown and the hundreds named. The town plan is particularly interesting in that Reading is not in Buckinghamshire and Speed explains, within the strapwork cartouche, that 'Barkshire could not contayn place for this Towne'.*

3a. *Lancashire by Pieter van den Keere, from the 'miniature Speed' atlas of 1627. The cartouche resembles fretwork but otherwise the map is plainly decorated. The long-vanished Marton Mere and Martin Mere (unnamed) can be clearly seen.*

3b. *One of the curious little maps which illustrated the eighteen songs by Michael Drayton in 'Poly-olbion' (1612). Little topographical detail is shown and the maps are covered with mythological and symbolical figures.*

4a. *Extravagantly ornamented cartouche from Saxton's Cornwall map (1579) engraved by Leonard Terwoort.*

4b. *Eighteenth-century decoration—vignette illustrating scenes of county life from the map of Nottingham in Emanuel Bowen and Thomas Kitchin's 'Large English Atlas' (1760).*

5. (a) Cartouche engraved by William Hole for the map of Lancashire in Camden's 'Britannia' (1607). It is copied from the cartouche by Hogenberg on Saxton's map of Lancashire. (b) Corner ornamentation from Ortelius's map of the Americas (1590). (c) Heraldry on the map of Essex by Joannes Blaeu (1645). (d) Decorative scale of miles from the map of Denbigh and Flint by Joannes Blaeu (1645).

6. *Eastern section of the map of Savoy by Joannes Jansson (1650). Here the unrivalled elegance of the work of Jansson is clearly demonstrated. The scales of French and Italian miles should be noted.*

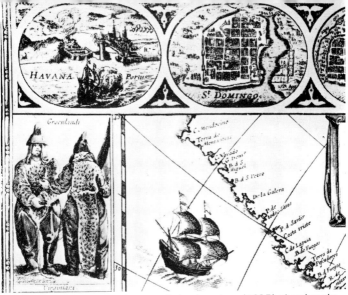

7a. *Detail from Blaeu's map of the Americas (1635) showing the use of town plans and local figures to form a decorative and useful border.*

7b. *Detail from Blaeu's map of West Africa (1635) showing nicely engraved fauna of the region.*

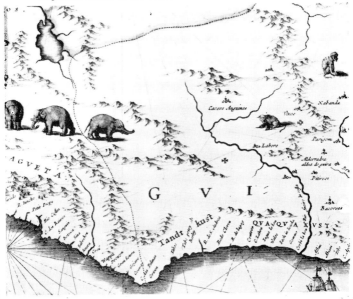

8. *Section of Blaeu's map of Denbigh and Flint (1645) with the most delicate engraving of topographical detail. Offa's Dyke is graphically depicted running in a south-easterly direction.*

9. *Section of Oxford from 'Britannia' by Richard Blome (1673). The crudity of the lettering and ornamentation is apparent when compared with the work of Blaeu opposite.*

10. 'The road from Nottingham to Grimsby'—a small detailed map from 'Britannia Depicta' (1720) by John Owen and Emanuel Bowen.

11. *Section of the road from York to W. Chester from Owen and Bowen's 'Britannia Depicta' (1720). The strip maps are reduced from those in Ogilby's 'Britannia' (1675).*

12. *Huntingdonshire from Hermann Moll's 'A New Description of England and Wales' (1724). Moll's county maps are plain and noteworthy chiefly for the antiquities illustrated in the margins.*

13. *Cambridgeshire (1750) from a series of county maps drawn by Thomas Kitchin for the 'Universal Magazine'. The cartouche incorporates the title on a scroll and uses figures from the academic and racing worlds to illustrate the county's life.*

NORTHAMPTON,
Divided into its
HUNDREDS;
Exhibiting the
City, Borough, Market Towns &c.
with various Historical Extracts relative to
Trade, Manufactures &c.
Describing also the Church Livings, with
Improvements not inserted in any other
Half Sheet Map Extant.
By Eman: Bowen,
Geographer to His late Majesty.
& Tho.s Bowen.

NORTHAMPTON SHIRE is Divided into 20 Hundreds, contains
250000 Acres of Land, 1+800 Houses, 326 Parishes, 11 Market Towns,
returns 9 Members to Parliament. Its 50 miles in length, 21 in breadth
and about 120 in Circumference. This County is of disproportional note
for its Number of Noblemen & Gentlemens Seats, having more
Parks than any other County in England. Nor is any County in
the Kingdom better stored with Grain, or has a better
race of Sheep, whose Wool has yielded a fund for the
Woollen Manufacture. The Hills which are neither high
nor barren, afford every where delightful Prospects. In
some places there are 30 and 40 Steeples in view at one
time. This County is also noted for a good breed of Horses.

NORTHAMPTON 68 miles from London is one of the
prettiest Towns in England. It was burnt down Sept. 20.
1675, but soon rebuilt again much finer and more uniform than
it was before. Tis a Borough Town, governed by a Mayor, Recorder,
a Bailiff, 2 Aldermen, & Common Council Men. Town Clerk &c. Tis
a principal Street, open to the 4 Cardinal Points. The Market place is esteem'd one
of the finest in Europe and its Market & Fairs celebrated for the best Horses in the
kingdom. The chief Manufactures are Shoes and Stockings, and the River Nen
being made Navigable, greatly contributes to the increase of its Trade
Here is an Infirmary for Sick People.

Rockingham is famous for a Castle and Forest and for giving a Title
of Honour to the Noble Family of the Watsons. The Forest is now
much less than formerly. Tis a great quantity of Charcoal is
still made and sent yearly to Peterborough

Near Naseby was fought the fatal
& decisive Battle between King
Charles the first and the Parliaments
Army, June 14.th 1645.

PETERBOROUGH 70 miles from London
Place of great Antiquity, but took its name
only from Church besides the Cathedral. It
here is not considerable the River Nen be-
Navigable renders the Traffic for the Impor-
of Coals and Corn and the Exportation of
other Goods, especially those of the Manufacture very convenient The Abbey
of the best endowed in England

P.t OF LINCOLN

PART OF
RUTLAND
SH.

PART OF
LEICESTER
SH.

Explanation.
Borough Towns with the Number of Members they
send to Parliament by Stars.
Rectories, Vicarages and Curacies.
Charity Schools. Religious Houses
The Figures on the Roads denote Measur of Dist.
between the Principal Towns.
The Market Days are annexed to their
respective Towns

P A R T   O F
HUNTINGD
SHIRE

P A R T   O F
BEDFORD SH.

Kettering is a populous Market Town, owing its prosperity to the Woollen Manufacture, but chiefly to Weaving & making of Shalloons, Serges and Tammies.

Higham Ferrers so call'd from the Ferrers who had antiently a Castle in
both healthy and pleasant. The Church which has a lofty Spire is a
some Structure It is govern'd by a Mayor, 7 Aldermen, a Recorder, a
petal Burgesses.   It sends only one Member to Parliament

B U C K I N G H A M

Brackley was once a famous Staple for
Wool, but is now noted for
its Hogs, Boots and Shoes.

Towcester is supposed to have been a Ro-
Town and Station The Inhabitants are employ'd
in the Manufacture of Lace and Silk.

B U C K I N G H A M

British Statute Miles 60 to 1 Degree

West 85. Long. 80 from London

15. Section of Derbyshire by Charles Smith in 'Smith's New English Atlas' (1804). The maps in this atlas are strikingly similar to those of Cary's 'New English Atlas' (1809) and are without decoration of any kind.

14. (opposite) Northamptonshire from 'Atlas Anglicanus' (1767) by Emanuel Bowen and Thomas Bowen. This is a reduced version of the larger maps in the 'Large' and 'Royal English Atlases' and well illustrates the use of copious descriptive notes relating to local topography, history and industry.

16. *Lancashire from 'The English Counties Delineated' (1836) by Thomas Moule. This series of maps represents a return to the spirit of decorative mapmaking. Canals are shown and the Manchester and Liverpool Railway with a branch to Bolton and a projected line southwards from Manchester.*

ments were used in 1740 for the setting up of milestones along main roads.

The surveyed material was assembled into a hundred plates, each bearing six or seven strips of road, each strip being $2\frac{1}{2}$in. wide (Fig. 9). The plates formed a folio volume which portrayed its information with striking clarity but was still too big and heavy to be carried by a traveller.

Ogilby explained the method of using the strip maps in his Preface . . . 'The Initial City or Town being always at the Bottom of the outmost Scroll on the Left Hand; whence your Road ascends to the Top of the said Scroll; then from the Bottom of the next scroll ascends again, thus constantly ascending till it terminate at the Top of the outmost Scroll on the Right hand . . . the Road itself is express'd by double Black Lines if included by Hedges, or Prick'd Lines if open . . . The Scale . . is according to one Inch to a Mile . . . the said Miles being exprest by double Points, and numbered by the Figures 1, 2, 3, &c. Each subdivided into 8 Furlongs, represented by Single Points. Ascents are noted as the Hills in Ordinary Maps. Descents e contra, with their Bases upwards'. This last convention looks odd indeed, exactly as if hills are drawn upside down in some places.

Britannia laid the foundations for a long series of British road-books and had far-reaching effects on future mapmaking as well as benefitting the traveller at a time when land travel was on the increase. From Ogilby's time onwards roads were included as a matter of course on all maps and so practical is the strip method that it is widely used in little-changed form today.

Robert Morden

In the 17th century the London map trade was largely concentrated in two areas—firstly in and around Cornhill, and secondly in St. Paul's Churchyard and the vicinity of Newgate and Cheapside. Robert Morden, a bookseller, publisher and cartographer, worked at the Atlas in Cornhill from 1668 to 1703 and was responsible for engraving county maps for Edmund Gibson's 1695 translation of Camden's *Britannia*. The preface to this edition of Camden states . . . 'The maps are all newly engrav'd, either according to Surveys never publish'd or according to such as have been made and printed since Saxton and Speed. Where actual Surveys could be had, they were purchas'd at any rate; and for the rest, one of the best copies extant was sent to some of the most knowing Gentlemen in each County, with a request to supply the defects, rectifie the positions, and correct the false spellings.

And that nothing might be wanting to render them as complete and accurate as might be, this whole business was committed to Mr. Robert Morden, a person of known abilities in these matters, who took care to revise them, to see the slips of the Engraver mended, and the corrections, return'd out of the several Counties, duly inserted. Upon the whole, we need not scruple to affirm, that they are by much the fairest and most correct of any that have yet appear'd'. Although Morden's maps incorporate several innovations they hardly live up to this confident panegyric. Neither in craftsmanship nor design do they compare with Blaeu or Jansson. On the credit side Morden undertook considerable revision of place-name spellings and many names appear on his maps as they do today. The Morden maps, being a post-Ogilby series, include a network of roads, some of which do not, in fact, appear in Ogilby. Strangely, however, Morden continued to use three scales to represent 'Great', 'Middle' and 'Small' miles, which appeared to be 2430, 2200 and 1830 yards respectively.

Following the precedent set by John Seller in his map of Hertfordshire (1676), Morden used London (St. Paul's and not, as later, Greenwich) as his prime meridian. His output was large and included a set of small county maps (4¾in. by 5⅞in.) in *The New Description and State of England* (1701) as well as a pack of playing card maps.

BRITISH REGIONAL MAPPING: 1700-1860

The two halves of the 18th century contrasted markedly as far as cartographic development is concerned. Before 1750 there was not much originality and little indication of those developments which were to occur later in the century—developments which established mapmaking in this country on a firm, scientific basis. In the early 1700s there was continued reworking of earlier material and further reprints of Saxton, Speed and others appeared. Such innovations as did occur related not so much to map detail as to secondary matters— the depiction of antiquities in the outer margins of Hermann Moll county maps of 1724; the use of copious descriptive notes on the face of Bowen and Kitchin maps; and the reduction of Ogilby's *Britannia* into pocket-sized road books by Bowen, Senex and Gardner.

The second half of the century, however, began propitiously with the Royal Society of Arts encouraging prospective surveyors with the award of an annual prize for a survey of a county at the, for those days, large scale of one-inch to one mile. This award sparked off a profusion of county surveys and thenceforward the one-inch scale was established as the standard scale it remains to this day. Apart from the publication of such large-scale county maps two other significant events occurred in the latter half of the 18th century. In 1791 the Ordnance Survey was established, followed by the founding of the Hydrographic Office in 1795, thus putting both map and chart making on an official footing, although private county surveys continued until well into the 19th century.

Ogilby 'Improved'—pocket road books

From the contemporary traveller's point of view, the disadvantage of Ogilby's *Britannia* was its bulk (it was a folio volume weighing $4\frac{1}{2}$lbs.) and it is surprising that over forty years were to elapse before anyone brought out a handier, pocket-sized road book. In 1719, however, John Senex, a London cartographer, engraver and publisher, issued *An Actual Survey of all the Principal Roads of England and Wales . . . first perform'd and published by John Ogilby, Esq.; and now improved . . . by John Senex.* The 'improvements' related only to the reduction in bulk, for the smaller scale of half-an-inch to one mile meant that some of Ogilby's detailed information had to be omitted. Some earlier charm was lost also, for Ogilby's cartouches with their rural scenes were replaced by plain title panels. The new volume contained one hundred plates, like Ogilby's, with six or seven strips to a page, the strips now reduced to one-and-a-quarter inches wide. In this same year Thomas Gardner published *A Pocket Guide for the English Traveller* which was another road book of one hundred strip maps reduced from Ogilby.

Emanuel Bowen, one of the most significant mapmakers of his century, issued a pocket book of roads in 1720, in collaboration with John Owen. This was entitled *Britannia Depicta or Ogilby Improv'd* and contained small county maps (Plate 10) as well as strip maps (Plate 11), the whole amounting to two hundred and seventy-three plates. With customary scorn for any false modesty, the title page vows . . . 'the Whole for its Compendious Variety & Exactness, preferable to all other Books of Roads hitherto Published or Proposed; and calculated not only for the direction of the Traveller (as they are) but the general use of the Gentleman and Tradesman'. Each set of strip maps for a particular road is preceded by a county map surmounted by a baroque cartouche enclosing title and a table of distances. Despite their small dimensions, the county maps convey much information—hundreds, rivers, hills, towns (with Parliamentary representation indicated by asterisks), churches and roads. The space between county boundary and frame is taken up with descriptive details relating to the geography and history of the county, a practice carried to great lengths in Bowen's later and larger atlases. *Britannia Depicta* proved popular and twelve further editions appeared up to 1764. The maps, which are not difficult to acquire today, are attractive items for the collector and look well when hand-coloured and framed.

Herman Moll (1688-1745)

Moll, a Dutchman who came to London in 1688, was a prolific publisher and engraver of maps and atlases, as well as an inveterate critic of his rivals. On his map of South America, other publishers are angrily dismissed as 'ignorant pretenders' and on his world map he says of his fellow countrymen—'as for ye Dutch maps all of 'em yet extant, are much alike and far enough from Correctness'. Moll, in truth, had little reason to criticise—his set of county maps in *A New Description of England and Wales* (1724) displayed little originality and no remarkable craftsmanship. They are, however, distinctive for the antiquities engraved in the margins (Plate 12). The Lancashire map, for instance, has illustrations of Roman relics discovered at Ribchester, Upholland and Standish. The map detail is derived from earlier sources such as Ogilby, Norden and Speed for there were still no copyright laws to prevent the plagiarism of other publishers' work. It was only in 1734 that the first copyright act was introduced in this country.

George Bickham and 'The British Monarchy'

In 1754 George Bickham (1684-1771) published *The British Monarchy,* a volume of one hundred and eighty-eight plates of historical notes with forty-three plates of views of English and Welsh counties. These can hardly be called maps for they are merely perspective sketches drawn from a suitable foreground eminence with the county stretching away into the distance. Much of each plate is occupied by a foreground scene showing, in Bickham's words '. . . a pleasing Landscape . . . with a variety of Rustic Figures, Ruins &c and the names of the Principal Towns and Villages, interspersed according to their apparent situation'. The detail shown is minimal—in the case of Westmorland only eleven towns, two rivers and 'Winander Meer' appear. The descriptive text of the volume is written in Bickham's elegant calligraphy for he was a celebrated writing master who wrote various books on the subject including *The Universal Penman.*

John Rocque

Some of the finest maps of the 18th century came from the hand of the Huguenot immigrant surveyor, John Rocque. Before 1750 Rocque was employed in preparing plans of great houses and during this time developed a strikingly effective way of engraving pasture, gardens, heath, cultivated land and other forms of land use. This talent is clearly shown on his fine county maps of Shropshire, Middlesex, Berkshire and Surrey and particularly on his remarkable large-scale plan of the Cities of London and Westminster published in 1746, a year in which he issued another fine plan of London and the country ten miles round. In addition to his unique differentiation between categories of land use, Rocque showed hills in plan with lighting from above so that the tops appeared white and lines were drawn down the slopes, thicker to indicate where the ground was steeper. Rocque issued a set of small county maps, 6in. by 7¾in., in *The English Traveller* (1746) which were re-issued in 1753, 1762 and 1764 in *The Small British Atlas*. Rocque had a large output, a contemporary advertisement of his work listing over seventy items, among which his plans of Bristol and Dublin stand out.

Literary periodicals

Around the middle of the 18th century, newly-established periodicals such as *The London Magazine; or Gentleman's Monthly Intelligencer* and *The Universal Magazine of Knowledge and Pleasure* were using cartographical illustrations. A set of county maps by Thomas Kitchin appeared in the former between 1747 and 1760 (Plate 13), while fifty-one maps by Emanuel Bowen, Thomas Kitchin and R. W. Seale were issued in *The Universal Magazine* between 1747 and 1766 (Fig. 10).

'The Large English Atlas' of Emanuel Bowen and Thomas Kitchin

Among the most interesting maps of the mid 18th century were those prepared by Emanuel Bowen and Thomas Kitchin for *The Large English Atlas* (1760). Bowen, an engraver and print seller, planned an atlas of county maps which were to be the largest and most detailed so far. His scheme eventually bore fruit in *The Large English Atlas* but his own monetary resources were insufficient to carry the project through alone, and he joined forces with Kitchin, another London engraver and publisher. They shared the engraving of the maps almost equally but Middlesex was the work of R. W. Seale and differs from the rest in that it incorporates the arms of the ninety-two City Livery Companies and of the City of London. Prior to the publication of the complete atlas the county maps were issued singly and consequently have differing dates of publication. Cornwall, for example, is dated 1750, Cheshire 1751 and Lancashire 1752. These were the largest county maps to appear up to their time (27in. by 20in.) but the scales vary with the size of the county and the way it fits the page. The inclusion on each map of a dedication to the Lord Lieutenant of the county, together with a list of the 'Seats of the nobility &c' indicates the promotors' awareness of the need for attracting local interest and subscriptions to help finance publication.

The most distinctive feature of Bowen and Kitchin's maps is the insertion of copious descriptive notes on the face of the map between boundary and frame (Plate 14). These relate to the topography and history of each county. Each map has a splendid cartouche or vignetted scene of the life of the county (Plate 4B). The titles invariably begin 'An Accurate Map of . . .' and a smaller Rococo cartouche contains the dedication. The borders are formed by thick outer lines with double inner lines marked off in degrees and minutes, while the maps themselves are covered by a graticule. Considerable topographical detail is included and the 'explanations' list boroughs, market towns with their market days, villages with R or V to indicate rectory or vicarage, charity schools, parks and post stages.

The 'Royal English Atlas'

Bowen and Kitchin, each of whom had a large output of maps and atlases, issued another fine atlas, the *Royal English Atlas,* in 1762. Smaller in format than the *Large English Atlas* it was designed on similar lines. The county maps were at a reduced scale and with less elaborate cartouches but still had a good deal of verbal description inserted wherever space allowed.

County surveys at a scale of one-inch to one mile

By mid 18th century, then, Bowen had attempted to encompass the whole country at a reasonably large scale. Individual counties had, however, been surveyed at a scale of one-inch to one mile and, of these, Henry Beighton's map of Warwickshire, surveyed 1722-25, is outstanding, partly because it was one of the first county maps to be soundly based on trigonometrical survey. Beighton, a meticulous and accomplished surveyor, illustrates the framework of triangles he used in making his survey in the bottom right-hand corner together with surveying instruments used, including a 'plotting-table' which he invented in 1721. A remarkably wide range of fea-

10. *A map of Lancashire engraved by Emanuel Bowen for the 'Universal Magazine', 1751.*

tures are depicted on the map, reflecting the whole life and economy of the county:— parish churches, chapels, depopulated places, seats of nobility, chases, parks, kings' houses, monasteries, castles, Roman ways and stations, battles, garrisons, coal mines, pumping engines, mills and medicinal waters. The inclusion of shields bearing subscribers' names is once again evidence of the practice of financing production by soliciting private subscriptions.

Beighton's splendid map was well ahead of its time and it was only after 1750 that much progress was made in mapping England at the one-inch scale. By the end of the century, however, the whole country, together with parts of Wales and Scotland had been mapped at this scale or larger.

The Royal Society of Arts award

The greatest stimulus for scientific mapmaking came from the Society for the Encouragement of Arts, Manufactures and Commerce (known since 1847 as the Royal Society of Arts). The Society's original objective was the raising of public funds with which to provide awards for meritorious discoveries and inventions in art, industry and commerce. In 1755 William Borlase submitted to his friend, Henry Baker, one of the most scientifically-minded members, a project for the Society's consideration in these terms . . . 'I would submit to you as a Friend whether the State of British Geography be not very low, and at present wholly destitute of any public encouragement. Our maps of England and its counties are extremely defective . . . if among your premiums for Drawings some reward were offered for the best Plan, Measurement and Actual Survey of City or District, it might move the Attention of the Public towards Geography, and in time, perhaps, incline the Administration to take this matter into their hands . . . and employ proper persons every year, from actual Surveys, to make accurate Maps of Districts, till the whole island is regularly surveyed'. This is a plea, therefore, not only for a national survey but also, with foresight, for regular revision of maps. For various reasons it was not acted upon until March 1759 when the Society published an advertisement offering an award of not more than £100 for an original survey upon a scale of one-inch to one mile. Between 1759 and 1801 twenty-three surveyors submitted claims and thirteen county maps were successful, the first being a map of Devonshire by Benjamin Donn. Two important points are made by Donn in the proposals for his map; first, 'As for the Accuracy, and consequently the value, of a map must chiefly depend on the Correctness of the Position, and horizontal Distances of the principal Places, particular care will be taken to determine these in a new and rational Method, by the Assistance of a curious set of Instruments, Trigonometrical Calculations, and Astronomical Observation' and second, 'The Roads, (at least the High Roads) will be actually measured'. Donn's map, therefore, was established on firm scientific foundations and the actual survey took five-and-a-half years in which over six thousand miles of roads and rivers were surveyed, together with the angles of towers and hills.

Notwithstanding the scientific construction of the new maps, the decorative arts were not entirely forgotten. Many maps were embellished with large and splendid cartouches. The map of Essex, for example, published in 1777 by John Chapman and Peter André—one of the most detailed county maps

—has a cartouche showing a fulling-mill with cloth beaters; Andrews and Dury's Wiltshire, published in 1773, includes a scene symbolising the life and products of the county with wheat, sheep, milking and a bale of cloth. This latter cartouche was specially designed by G. B. Cipriani and engraved by T. Caldwell, the practice of engaging a special engraver solely for a cartouche being fairly common at this time.

An outstanding figure in mid 18th century cartography was Thomas Jefferys, Geographer to George III and a noted authority on North America. Besides engraving Donn's prize-winning map of Devon, Jefferys was himself responsible for publishing maps of Bedford, Huntingdon, Buckingham, Cumberland, Oxford, Durham, Westmorland and a fine twenty-sheet map of Yorkshire. He depicted relief by hachuring (finely-engraved lines drawn down the slopes, thick where the slope was steeper and thinner where it was more gradual) but while this brought out prominent hills and valleys clearly, it was less successful in portraying a general picture of the topography.

Several of the new large-scale maps included inset plans of principal towns or views of prominent buildings. Isaac Taylor's splendid map of Gloucestershire (1777) has engravings of castles, Chapman and André's Essex includes a plan of Colchester and Jefferys' Buckinghamshire has an inset of Buckingham itself.

Various surveyors were at work throughout the country, some producing maps of several counties. Andrew Armstrong produced maps of Northumberland (a winner of the Royal Society award), Durham, Lincoln and Rutland; William Yates, a Liverpool customs officer, surveyed Lancashire, Staffordshire and Warwickshire; Peter Burdett, a Derbyshire artist, made maps of Cheshire and Derbyshire; and others such as Joseph Lindley, J. Prior, W. Day, T. Milne, J. Hodskinson, J. Evans and T. Eyre mapped individual counties. In some cases the surveyors provided evidence of the scientific framework of their maps in the form of triangulation diagrams. The triangulation system used in county maps of the Midlands and north-western England can be said to fairly anticipate the principal triangulation of the Ordnance Survey and the fact that the same stations have been selected, in many cases, by both private and official surveys is a pointer to the sound concepts on which the private surveys were founded.

Maps were still issued in black and white only with hand-colouring added to order. Publication was in the form of single sheets and the catalogue of W. Faden, who took over

Jefferys' business in Charing Cross and bought up many copper plates of 18th century surveys, gives some measure of map prices in the early 19th century. Yates' Lancashire cost £1 12s., Isaac Taylor's four-sheet map of Hereford was 16s., Rocque's eighteen-sheet map of Berkshire was £1 12s. The average size of an edition of a county map would be about three hundred copies, of which most would go to subscribers[1]. An extra charge of five shillings would be made for hand-colouring and a specially-bound edition with title page and printed index such as might be found in the libraries of county gentry would cost a further two guineas.

John Cary (c. 1754-1835)

Despite the domination of this period by large-scale individual county maps, some publishers, of whom John Cary and Charles Smith were outstanding, were producing small, reasonably-priced atlases which represented an improvement on anything which had appeared so far.

The cartographic historian, Sir H. G. Fordham, regarded Cary as the 'most representative, able and prolific of English cartographers'—high praise from a man who had made a special study of Cary's life and work. Whether we agree or not with this assessment, it cannot be denied that Cary was a fine craftsman who had a formidable publication list. His *New and Correct English Atlas* (1787) was a quarto volume of forty-six county maps, each accompanied by descriptive text, which were an advance on any comparable set of maps issued to date for Cary was able to draw freely on the work of larger-scale county surveyors. For Gough's translation of Camden's *Britannia* (1789) Cary engraved fine county maps, 16¼in. by 18¼in., which were characterised by clear engraving and included considerable detail including hundred boundaries (Fig. 11). These maps were drawn by E. Noble and engraved by Cary. The *Traveller's Companion* (1789) was an octavo volume with forty maps of English counties, a general map of England and Wales, and maps of north and south Wales, all printed on one side only of thin paper. At the top of each map is a title panel with the county name lettered against a hatched background. On the left of the panel are the words 'By J. Cary' and on the right 'Engraver'. The imprint 'London. Published Sepr 1, 1789 by J. Cary, Engraver No. 188 Strand' appears below the maps. In spite of their small dimensions, only 3½in. by 4¾in., these maps are clear and legible with those turnpike roads traversed by mail coaches in blue and others

[1] *The Large Scale County Maps of the British Isles, 1596-1850,* Elizabeth M. Rodger, Bodleian Library, Oxford, 2nd edition, 1972.

11. Western section of John Cary's map of South Wales in Gough's edition of Camden's 'Britannia'. Hills are crudely depicted in plan with lines drawn down the slopes and with light falling directly on to the summits which are left white. The map is bare of decoration.

red. A bold type picks out the names of market towns and distances are indicated along the roads. The *Traveller's Companion* proved deservedly popular and was re-printed several times up to 1828.

Cary's finest series of maps appeared in his *New English Atlas* (1809). The title of each begins 'A New Map of. . . .' and ends 'By John Cary Engraver' and all have the imprint 'London: Published by J. Cary, Engraver & Mapseller No.

181 Strand' together with the date. These were Cary's largest county maps, measuring 18¼in. by 20¼in. The atlas, one of the finest products of 19th century mapmaking, was beautifully engraved and, though functional, the maps are quite attractive when hand-coloured. The *New English Atlas* was issued in parts between 1801 and 1809 so that counties have varying dates of publication. Many are dated 'Sepr. 28, 1801' and others vary from December 21, 1801 to June 1, 1809.

Roads appeared prominently on Cary's maps for he was employed by the Postmaster-General to organise the survey of turnpike roads in Great Britain, a task involving nine thousand miles of survey. Decoration is eschewed on all the Cary series and this assists his intent to present geographical information clearly and without distraction.

After the closing of the Cary firm c. 1850, G. F. Cruchley took over the plates and used them for future editions, updating them regarding railways and parliamentary divisions.

Charles Smith

Cary's competitor, Charles Smith, published a *New English Atlas* in 1804 well before Cary's work of the same title. The two sets of maps are so similar in style and conception that there has been speculation as to whether one or the other had copied his rival's work (Plate 15). An alternative explanation could be that both worked from common sources and therefore it was hardly surprising that they produced factually similar maps. The visual similarity, on the other hand, is not so readily explained—it seems unlikely that a man of Cary's calibre, who had already published a fine atlas in 1787, should stoop to such plagiarism. Nevertheless the fact remains that each man produced work of exceptional merit which is not disgraced when compared with the early one-inch to one mile sheets of the Ordnance Survey which were appearing slowly from 1801 onwards.

Private one-inch scale county mapmaking

The development of the official mapmaking body, the Ordnance Survey, did not as yet cause the demise of the private county surveyor. Men like Christopher Greenwood had their own projects for one-inch scale coverage of the whole nation and, indeed, they were assisted to some extent by the Trigonometrical Survey of the Board of Ordnance, as it was then called, for one of its functions in those early days was to provide private mapmakers with the locations of trigonometrical points.

Christopher Greenwood

A Yorkshireman, Christopher Greenwood, and the firm of which he was head played a prominent role in early 19th century mapmaking. Although his scheme for a national set of one-inch maps did not come fully to fruition, he almost did succeed. After an initial survey of his native county in 1817-18, Greenwood published maps of thirty-four other counties, leaving only six English counties and three in Wales still to be surveyed. His method of promotion was to insert a prospectus in the local press outlining the objectives and methods of his proposed scheme and to follow this with a more detailed prospectus which was sent to advance subscribers. An important part of his sales technique was to publish a list of subscribers, headed by nobility but including customers from the business and professional classes. Vignettes of buildings and landscape scenes appeared on a number of his maps and, in his prospectus, Greenwood wrote . . . 'with a view to render the County Maps as ornamental as useful, the Proprietors will use every means to join superior elegance with minutest accuracy. Vignettes from the pencils of distinguished artists will be added to the Maps of such Counties as furnish appropriate subjects'. Richard Creighton's meticulous engraving of Durham Cathedral on the Durham map is an outstanding example of these vignettes.

The map of Worcestershire is typical of the Greenwood maps—a large map in four sheets engraved by Neele & Son (Fig. 12). The eye is at once drawn to the flamboyant title and dedication which are a curious amalgam of numerous styles of lettering. This title, *Map of the County of Worcester from Actual Survey made in the Years 1820 & 1821 by C & J Greenwood, London. Published by the Proprietors G. Pringle Junr & C. Greenwood, 70 Queen Street, Cheapside. 1st June 1822. Engraved by Neele & Son, 352 Strand,* uses no less than ten different types of alphabet and this is characteristic of Greenwood's maps. The border is formed of a thick line between two fine ones, with a double inner line marked off into degrees and minutes of latitude and longitude. Between the two sets of lines is a hatching with double lines at the outer edge resembling piano keys. This type of border is typical of the time and may have been derived from early Ordnance Survey sheets. The maps show considerable human and topographical detail with relief shown, not very satisfactorily, by fine hachuring.

The firm was known as C. & J. Greenwood, Christopher having his younger brother John as partner along with G. Pringle and G. Pringle Junior. After 1828 when the business was failing, Greenwood devised two schemes to help bolster the finances. The first was for an atlas of county maps at a scale of approximately three miles to one inch. This was issued in four parts with excellent maps, similar in design to the larger-scale series, and engraved by well-known craftsmen. Greenwood himself was not an engraver and the cost of engraving must have been a heavy drain on the firm's resources. Greenwood's second and final scheme, which failed to materialise, was for the establishment of a body which would undertake regular revision of the county maps in the Greenwood atlas—an idea reminiscent of that put to the Royal Society of Arts by William Borlase many years earlier. On the final collapse of the Greenwood firm, some copper plates were purchased by other publishers, among them Henry Teesdale whose company itself published a notable county atlas, the *New British Atlas* (1829).

Andrew Bryant

Another surveyor with ideas of mapping the whole country was Andrew Bryant and his maps were generally more detailed than Greenwood's and at a larger scale—one-and-a-half inches to one mile. Bryant succeeded in mapping only thirteen counties but, in some cases, there was keen rivalry—both men published maps of Surrey in 1823 and Gloucestershire in 1824 —and at times it appears that Bryant, Greenwood and the Ordnance Survey must have been working simultaneously in the same county. Both Greenwood and Bryant's maps were available in various forms—flat, mounted on common roller, mounted on spring roller, folded in case, in case half-bound, coloured or uncoloured—and at prices varying with the luxury of the presentation. The standard subscription for Greenwood's projected series of one-inch maps was one hundred and twenty-five guineas or three guineas per map.

Small-scale 19th century atlases

The one-inch scale did not entirely monopolise the 19th century and county atlases of varying size and quality continued to appear, as well as maps for road books and directories. *Wallis's New Pocket Edition of the English Counties or Traveller's Companion* (1810) containing forty-four county maps, each 4in. by $5\frac{1}{2}$in., with descriptive text was typical of the period. In 1829, however, we find a significant development in Pigot & Company's *British Atlas* whose title tells us of the important fact that . . . 'the whole was engraved on steel plates'. The great advantage of steel for

12. *Detail from the one-inch to one mile map of Worcestershire by C & J Greenwood, 1822, showing a small detached portion of the county.*

engraving, compared with copper, was its greater durability which allowed much longer printing runs.

In 1836 Thomas Moule, a scholar with antiquarian interests, published a set of county maps in *The English Counties Delineated* which are popular with collectors because of their return to the tradition of decorative cartography. The maps, which are small but clearly engraved on steel, have decorative borders, heraldry, title cartouches and attractive vignettes of county scenes (Plate 16). They are easily obtainable and reasonably priced and look especially pleasing when hand-coloured.

J. & C. Walker, who engraved many plates for Greenwood, published their *British Atlas* in 1837 with detailed county maps which later formed the basis for a 19th century cartographic curiosity, *Hobson's Fox-Hunting Atlas* (1850), in which the areas and meeting places of the hunts were super-imposed.

Many 19th century atlases can be found easily today and are, naturally, lower priced than earlier and scarcer material. They can, moreover, be of interest to the local historian for their identification of roads, canals and railways. A word of caution is necessary, however, about taking such information at face value. Some of these maps are notoriously unreliable and whenever possible it is desirable to check their information with alternative sources.

THE FORMATIVE YEARS OF THE ORDNANCE SURVEY

Although the official beginnings of the Survey date from 1791 the impetus which pushed forward its foundation can be traced to the last Jacobite rising when Prince Charles Edward, at the head of his highland army, marched as far south as Derby. Lacking support in England he retreated over the border to be defeated at Culloden in 1746. The subsequent pacification carried out by the Duke of Cumberland in the highlands was hindered for want of good maps and a dual programme of mapping the highlands and opening them up with military roads was instituted. The roads were to be built under General Wade's supervision and their construction would permit rapid movement of troops in the event of further Jacobite excursions. The task of mapping originated with Colonel Watson and this survey may be regarded as the real birth of the Ordnance Survey. Work began in 1747 and the principal name associated with it is that of William Roy. The survey was carried out with compasses traverses, detail being filled in by field sketching, and these historic maps are now stored in the British Museum. In 1765 Roy was made Surveyor General of Coasts and Engineer for making and directing Military Surveys under the honourable Board of Ordnance, thus establishing a firm connection between the Board of Ordnance and mapmaking. The pursuance of the War of American Independence hindered progress towards a national survey but in 1783 a new initiative came from French astronomer, M. Cassini de Thury, who argued the scientific advantages of linking the observatories of Paris and Greenwich by precisely surveyed triangulation. His suggestion was accepted on this side of the Channel and Roy placed in charge of operations, his task being to measure a baseline on Hounslow Heath from which a triangulation system would reach out, linking Greenwich to Dover and subsequently dovetailing with the French triangulation. The baseline measurement was completed in 1784 and, after Roy's death in 1790, triangulation was continued by the Duke of Richmond, Master of the Ordnance. In 1791 Richmond brought surveying further under the wing of the Ordnance by establishing the Trigonometrical Survey with headquarters alongside those of the Ordnance in the Tower of London. The aims of this new body were to establish an accurate trigonometrical framework for the whole country and to produce a series of one-inch maps. The first map, a four-sheet map of Kent, appeared in 1801 and the next four were of Essex. Like their successors for some time to come they were engraved on copper, allowing thin lines to be clearly printed, as may be well seen if the hachuring on these maps is studied. Alterations to the plates were easy, for the part to be amended had only to be scraped, ground and burnished before re-engraving. These early copper plates measured 36in. by 24in. and weighed c. 35lbs. Before engraving a draughtsman would prepare a drawing on fine card for the engraver, reducing the larger-scale survey work to the one-inch scale. (The survey was mainly at two inches to one mile but in some areas at three and six inches to one mile.) Roads and lettering would be correctly drawn in and hachuring added with broad strokes. Progress was slow, a single sheet taking months to prepare, but the result was a clear, comprehensive study from which the engraver could proceed at the reduced scale.

The Ordnance maps were an improvement over many late 18th century county maps but there was, nevertheless, after 1820 some criticism as to their accuracy. This may be partly explained by the stringency of the Survey's budget—from 1791 to 1811 a mere £52,000—which may have influenced the surveyors in their attitude to the relative importance of the trigonometrical and topographical stages of the survey for, certainly, their over-riding interest lay with the trigonometrical construction rather than in the detailed topographic infilling.

William Mudge directed the early stages of the survey and was succeeded in 1820 by Captain Colby, under whose directorship the one-inch map took on a lighter and more delicate look with the finest of hachuring and small, neat lettering. By 1840 only Scotland and the six northern counties of England remained unmapped at the one-inch scale. England and Wales south of the Hull-Preston line were covered by sheets 1 to 90 of the *Old Series*. North of this line the sheets of the *Old Series* were reduced from larger-scale (six-inch) surveys but were issued with numbers following on from those of southern England, i.e. 91 to 110. Each sheet number was divided into quarters as the maps were issued as quarter sheets so that we find 91NW, 91SW, 91NE, 91NW and so on. This large-sheet numbering was superseded in 1872 when it was decided to carry the six-inch survey southwards. The original quarter sheets were gradually withdrawn and re-numbered to form sheets 1 to 73 in the *New Series*. In this instance, however, we are not concerned with the development of the Ordnance Survey beyond the completion of the *Old Series,* sometimes loosely called the *First Edition.* This had taken seventy years to complete but in 1870 the country had a scientific triangulation system and a series of maps rivalling any to be found elsewhere.

No table of conventional signs was issued with the *Old Series* though many symbols are used (Fig. 13). Turnpike and secondary roads are clearly shown with an indication of whether they were fenced or unfenced; tracks are given a double, dotted line but footpaths are shown only intermittently. Buildings in villages are solid black but in towns the built-up area is tinted or hatched. Windmills are delightfully drawn but, unike some of the county surveys, there is no symbol for watermills. Gardens are shown on the southern sheets but omitted on the later northern sheets. Occasional public houses are named, such as the Old Lamb at Kingston Bagpuize and on certain sheets hotels of interest to travellers, such as the Victoria at Llanberis, are shown. Unlike our more impersonal age, it was customary to credit the craftsmen and Ebenezer Bourne is constantly credited with the engraving of lettering while Benjamin Baker and his assistants engraved the hills. The maps are surrounded by a 'piano-key' frame with the scale and imprint usually placed below.

Though lacking consistency and with no real uniformity

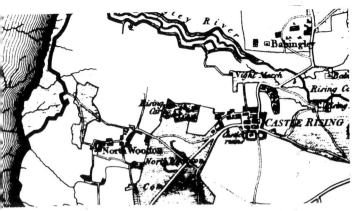

13. Detail from Ordnance Survey 'Old Series' one-inch to one mile sheet 69 (1824) showing parts of Norfolk and The Wash.

of style, the *Old Series* is a remarkable achievement, the first in a succession of Ordnance Survey series of the one-inch map which culminate in the splendid contemporary *Seventh Series*.

SUGGESTIONS FOR FURTHER READING

The following general books are recommended to the amateur enthusiast:

Bagrow, Leo. *History of Cartography*. Watts, 1964.

Baynton-Williams, R. *Investing in Maps*. Barrie & Rockliff, 1969.

Beresiner, Yasha. *British County Maps. Reference and Price Guide*. Antique Collectors' Club, 1983.

Hodgkiss, A. G. *Understanding Maps: A Systematic History of Their Use and Development*. Dawson, 1981.

Lister, Raymond. *How to Identify Old Maps and Globes*. Bell, 1965.

Lister, Raymond. *Antique Maps and Their Cartographers*. Bell, 1970.

Moreland, Carl, and Bannister, David. *Antique Maps: A Collector's Handbook*. Longman, 1983.

Radford, P. J. *Antique Maps*. Garnstone Press, 1971.

Tooley, R. V. *Maps and Mapmakers*. Batsford, fourth edition 1970.

Tooley, R. V., and Bricker, C. *History of Cartography — 2500 Years of Maps and Mapmakers*. Thames & Hudson, 1969.

Serious students and collectors of county maps and atlases will find the following works of inestimable value:

Hodgkiss, A. G., and Tatham, A. F. *Keyguide to Information Sources in Cartography*. Mansell, 1986.

Hodson, Donald. *County Atlases of the British Isles Published after 1703*. Volume I. *Atlases Published 1703 — 1743*. The Tewin Press, 1984.

Hodson, Donald. *County Atlases of the British Isles Published after 1703*. Volume II. *Atlases Published 1743 — 1763*. The Tewin Press, 1989.

Scott, Valerie G., and McLaughlin, Eve. *County Maps and Histories: Berkshire*. Quiller Press, 1984.

Scott, Valerie G., and McLaughlin, Eve. *County Maps and Histories: Buckinghamshire*. Quiller Press, 1984.

Scott, Valerie G., and Rook, Tony. *County Maps and Histories: Hertfordshire*. Quiller Press, 1989.

Scott, Valerie G., and Barty-King, Hugh. *County Maps and Histories: Sussex*. Quiller Press, 1985.

Shirley, Rodney W. *Early Printed Maps of the British Isles 1477-1650*. Holland Press, 1980.

Shirley, Rodney W. *The Mapping of the World — Early Printed World Maps 1472-1700*. Holland Press, 1983.

Skelton, R. A. *County Atlases of the British Isles*. Carta Press, 1970.

Smith, David. *Antique Maps of the British Isles*. Batsford, 1982.

Wallis, Helen, and Robinson, Arthur. *Cartographical Innovations: A Handbook of Mapping Terms to 1900*. Map Collector Publications, 1987.

Wilford, John Noble. *The Mapmakers. The Story of the Great Pioneers in Cartography — from Antiquity to the Space Age*. Junction Books, 1981.

The Map Collector is a finely illustrated quarterly journal entirely devoted to the interests of collectors of antique maps. It is obtainable from Map Collector Publications Limited, 44 High Street, Tring, Hertfordshire HP23 5BH. Telephone: 044282 4977.

WHERE TO OBTAIN AND CONSULT EARLY MAPS

Important collections are housed in the Map Library of the British Library, Royal Geographical Society, National Maritime Museum, Bodleian Library, National Library of Scotland and the Cambridge University Library. With the exception of the Royal Geographical Society it is necessary to obtain special permission to consult maps in these establishments. County Record Offices house map collections, both printed and in manuscript, related to their own county, while maps of local interest may be found in the libraries of larger towns and cities. The following dealers specialise in antique maps:

Antique Maps and Prints, 30 St Mary's Street, Stamford, Lincolnshire PE9 2DL. Telephone: 0780 52330.

Baynton-Williams, 37A High Street, Arundel, West Sussex BN18 9AG. Telephone: 0903 883588.

Billson of St Andrews, 15 Greyfriars Garden, St Andrews, Fife KY16 9HG. Telephone and fax: 0334 75063.

Brobury House Gallery, Brobury, Herefordshire HR3 6BS. Telephone: 09817 229.

Clive A. Burden Ltd, 92 Lower Sloane Street, London SW1W 8DA. Telephone: 071-823 5053.

Clive A. Burden Ltd, 46 Talbot Road, Rickmansworth, Hertfordshire WD3 1HE. Telephone: 0923 778097.

Carson Clark Gallery (Scotia Maps), 173 Canongate, The Royal Mile, Edinburgh EH8 8BN. Telephone: 031-556 4710.

Cartographia London, Pied Bull Yard, 18 Bury Place, Bloomsbury, London WC1A 2JR. Telephone: 071-404 4050.

Clevedon Books, 14 Woodside Road, Clevedon, Avon BS21 7JY. Telephone: 0272 872304. (By appointment only.)

Corner Shop, 5 St John's Place, Hay-on-Wye, Herefordshire. Telephone: 0497 820045.

Michael and Verna Cox, 139 Norwich Road, Wymondham, Norfolk NR18 0SJ. Telephone: 0953 605948. (By appointment only.)

Susanna Fisher, Spencer, Upham, Southampton SO3 1JD. Telephone and fax: 04896 291.

J. A. L. Franks Ltd, 7 New Oxford Street, London WC1A 1BA. Telephone: 071-405 0274; fax: 430-1259.

Frontispiece, 40 Porters Walk, Tobacco Dock, London E1 9SF. Telephone: 071-702 1678.

The Gallery, High Street, Yarmouth, Isle of Wight PO41 0PN. Telephone: 0983 760784.

Harrington Brothers, The Chelsea Antique Market, 253 King's Road, Chelsea, London SW3 5EL. Telephone: 071-352 1720. *Also at* Old Church Galleries, 320 King's Road, Chelsea, London SW3 5EP. Telephone: 071-351 4649.

Julia Holmes, South Gardens Cottage, South Harting, near Petersfield, Hampshire GU31 5QJ. Telephone: 0730 825040. (By appointment only.)

Hughes and Smeeth Ltd, 1 Gosport Street, Lymington, Hampshire SO41 9BG. Telephone: 0590 676324.

J. Alan Hulme, 52 Mount Way, Waverton, Chester CH3 7QF.

Telephone: 0244 336472.

Ingol Maps and Prints, Cantsfield House, 206 Tag Lane, Ingol, Preston, Lancashire PR2 3TX. Telephone: 0772 724769. (Postal business only.)

King's Court Galleries, 54 West Street, Dorking, Surrey RH4 1BS. Telephone: 0306 881757; fax: 75305.

Michael Lewis Gallery and Bookshop, 17 High Street, Bruton, Somerset BA10 0AB. Telephone: 074981 3557.

Lawson Gallery, 7 Kings Parade, Cambridge CB2 1SJ. Telephone: 0223 313970.

Leycester Map Galleries Ltd, Well House, Arnesby, Leicester LE8 3WJ. Telephone: 0533 478462; fax: 478268.

Maggs Bros Ltd, 50 Berkeley Square, London W1X 6EL. Telephone: 071-493 7160; fax: 499 2007.

Map House, 54 Beauchamp Place, Knightsbridge, London SW3 1NY. Telephone: 071-589 4325 or 584 8559, fax: 589 1041.

Richard Nicholson of Chester. Shop: 25 Watergate Street, Chester CH1 2LB. Correspondence to: Stoneydale, Christleton, Chester CH3 7AG. Telephone: 0244 26818 (shop) or 336004.

Avril Noble, 2 Southampton Street, Covent Garden, London WC2E 7HA. Telephone: 071-240 1970.

Oldfield Gallery, 76 Elm Grove, Southsea, Hampshire PO5 1LN. Telephone and fax: 0705 838042.

O'Shea Gallery, 89 Lower Sloane Street, London SW1W 8DA. Telephone: 071-730 0081; fax: 1386.

Powerhouse Gallery, 1 Market Lane, Laugharne, Dyfed SA33 4SB. Telephone: 0994 427635; fax: 427213.

Printed Page, 2/3 Bridge Street, Winchester, Hampshire SO23 9BH. Telephone: 0962 854072; fax: 862995.

Louise Ross and Co Ltd, Mulberry House, 8 Mount Road, Lansdown, Bath, Avon BA1 5PW. Telephone: 0225 448786; fax: 448789.

Sanders of Oxford Ltd, 104 High Street, Oxford OX1 4BW. Telephone: 0865 242590.

Henry Sotheran Ltd, 2-5 Sackville Street, London W1X 2DP. Telephone: 071-734 1150 or 0308.

Tooley Adams and Co Ltd, 13 Cecil Court, Charing Cross Road, London WC2N 4EZ. Telephone: 071-240 4406.

Robert Vaughan Antiquarian Booksellers, 20 Chapel Street, Stratford-upon-Avon, Warwickshire CV37 6EP. Telephone: 0789 205312.

Warwick Leadley Gallery, 5 Nelson Road, Greenwich, London SE10 9JB. Telephone: 081-858 0317.

Society of Map Collectors

The International Map Collectors' Society is a British organisation which is entirely concerned with the world of antique maps. Regular meetings are held in different parts of the UK and the world and an *IMCoS Journal*, published quarterly, contains articles on early maps as well as news of current events, reports on meetings and so on. Information about the Society can be obtained from the General Secretary, W. H. S. Pearce, 29 Mount Ephraim Road, Streatham, London SW16 1NQ. Telephone and fax: 081-769 5041.

The Antiquarian Map and Print Fair is held monthly at the Bonnington Hotel, 92 Southampton Row, London WC1B 4BH. Telephone: 071-242 2828.

Acknowledgments

I am much indebted to Dr J. B. Harley for his kindness in lending his expert knowledge to the reading of my manuscript; to D. H. Birch for his splendid photography of original maps; to Dr R. M. Prothero for permitting me to photograph a section of his Blaeu map of Guinea; and to Professor R. Lawton for allowing me to reproduce maps from the collection of the Department of Geography, University of Liverpool; other illustrations are reproduced from maps in my own collection.

INDEX

André, Peter 56
Andrews, J. 57
Anonymous Series (1602) 26
Apian, Philipp 13
Armstrong, Andrew 57
Babylonians 15
Baroque style 7
Beighton, Henry 54
Berey, Nicholas 9
Bickham, George 52
bird's eye views 12, 27
Blaeu family 8, 11, 20, 28, 29
Blome, Richard 30
Bodleian or Gough map 12, 17
borders 8
Bowen, Emanuel 8, 50, 51, 53, 54
Braun and Hogenberg 13
Bryant, Andrew 62
Buache, Philip 12
Burdett, Peter 57
calligraphy 10
Camden, William 26
Carte Pisane 18
cartouche 7
Cary, John 8, 58
Cassini de Thury 65
Catalan Atlas 18
Catalan mapmakers 18
Chapman, John 56
colouring 9
contours 12
copper engraving 22, 23
Day, W. 57
de Hooghe, Cornelis 22
de la Cosa, Juan 18
Donn, Benjamin 56
Drayton, Michael 28
Dury, A. 57
Ebstorf map 17
Eratosthenes 15
Evans, J. 57
Eyre, T. 57
Faden, W. 57
form lines 11
Gardner, Thomas 50, 51

Gastaldi, Giacomo 19
Greenwood, C. & J. 61
Harris, John 8
Hereford map 17
hill shading 12
Hodskinson, J. 57
Hogenberg, Remigius 22
Hondius, Jodocus 11, 20, 26, 27
Humble, George 27
impression 5
Italian school 8, 19
Jaillot, Alexis Hubert 21
Jansson, Joannes 11, 20, 30
Jefferys, Thomas 57
Jenner, Thomas 28
Kitchin, Thomas 50, 53, 54
Lafreri atlases 19
Latin terms 5
Lily, George 21
Lindley, Joseph 57
literary periodicals 53
lithography 5
Mercator, Gerardus 6, 20
mile, statue 9, 32
Milne, T. 57
Moll, Hermann 52
Morden, Robert 9, 49
Morgan, William 14
Moule, Thomas 64
Münster, Sebastian 10, 19
Norden, John 6, 14, 24-6
Nowell, Laurence 21
Nuremburg Chronicle 19
Ogilby, John 14, 30-2, 49
Old Series, Ordnance Survey 65, 66
Ordnance Survey 51, 61, 65-7
orientation 9
Ortelius, Abraham 6, 9, 20
Owen, John 51
Paris, Matthew 16
Pigot & Co. 62
Posidonius 15

Portolan charts 17
Prior, J. 57
Ptolemy, Claudius 15, 18, 19
reproductions 68
Reynolds, Nicholas 23
Ribero, Diego 18
road books 32, 51
Rococo ornament 8
Rocque, John 53, 58
Roy, William 64-5
Royal Society of Arts Award 50, 56
Ryther, Augustine 23
Sanson, Nicholas 20
Saxton, Christopher 22-4
scale 9
Scatter, Francis 23
Schedel, Hartmann 19
sea areas 10
Seale, R. W. 53
Seckford, Thomas 22, 24
Seller, John 50
Senex, John 50, 51
Smith, Charles 60
Smith, William 14, 26
Speed, John 6, 8, 14, 26-8
state of a map 5
Symonson, Philip 13, 24
Taylor, Isaac 57, 58
Teesdale, Henry 62
terminology 5
Terwoort, Leonard 22
T-O maps 16
towns 12
van den Keere, Pieter 28
Visscher, C. J. 20
Waghenaer, Lucas Jansz 20
Waldseemüller, Martin 18
Walker, J. & C. 64
Wohlgemut, Michael 19
wood engraving 19
Yates, William 57, 58